Lectures in Applied Mathematics

Proceedings of the Summer Seminar, Boulder, Colorado, 1960

VOLUME 1 LECTURES IN STATISTICAL MECHANICS
G. E. Uhlenbeck and G. W. Ford with E. W. Montroll

VOLUME 2 MATHEMATICAL PROBLEMS OF RELATIVISTIC PHYSICS
I. E. Segal with G. W. Mackey

VOLUME 3 PERTURBATION OF SPECTRA IN HILBERT SPACE
K. O. Friedrichs

VOLUME 4 QUANTUM MECHANICS
R. Jost

Proceedings of the Summer Seminar, Ithaca, New York, 1963

VOLUME 5 SPACE MATHEMATICS, PART 1
J. Barkley Rosser, Editor

VOLUME 6 SPACE MATHEMATICS, PART 2
J. Barkley Rosser, Editor

VOLUME 7 SPACE MATHEMATICS, PART 3
J. Barkley Rosser, Editor

Proceedings of the Summer Seminar, Ithaca, New York, 1965

VOLUME 8 RELATIVITY THEORY AND ASTROPHYSICS
1. RELATIVITY AND COSMOLOGY
Jürgen Ehlers, Editor

VOLUME 9 RELATIVİTY THEORY AND ASTROPHYSICS
2. GALACTIC STRUCTURE
Jürgen Ehlers, Editor

VOLUME 10 RELATIVITY THEORY AND ASTROPHYSICS
3. STELLAR STRUCTURE
Jürgen Ehlers, Editor

Proceedings of the Summer Seminar, Stanford, California, 1967

VOLUME 11 MATHEMATICS OF THE DECISION SCIENCES, PART 1
George B. Dantzig and Arthur F. Veinott, Jr., Editors

VOLUME 12 MATHEMATICS OF THE DECISION SCIENCES, PART 2
George V. Dantzig and Arthur F. Veinott, Jr., Editors

Proceedings of the Summer Seminar, Troy, New York, 1970

VOLUME 13 MATHEMATICAL PROBLEMS IN THE GEOPHYSICAL SCIENCES
1. GEOPHYSICAL FLUID DYNAMICS
William H. Reid, Editor

VOLUME 14 MATHEMATICAL PROBLEMS IN THE GEOPHYSICAL SCIENCES
2. INVERSE PROBLEMS, DYNAMO THEORY, AND TIDES
William H. Reid, Editor

Proceedings of the Summer Seminar, Potsdam, New York, 1972

VOLUME 15 NONLINEAR WAVE MOTION
Alan C. Newell, Editor

Volume 15

Lectures in Applied Mathematics

Nonlinear Wave Motion

Alan C. Newell, Editor

1974

American Mathematical Society, Providence, Rhode Island

The proceedings of the Summer Seminar were prepared by the American Mathematical Society with partial support from the following contracts and grants:

National Science Foundation under NSF grant GP-34254,
Office of Naval Research under contract N00014-73-C0015.

Library of Congress Cataloging in Publication Data CIP

Main entry under title:

Nonlinear wave motion.

(Lectures in applied mathematics, v. 15)
Proceedings of the Summer seminar, sponsored by the American Mathematical Society and the Society for Industrial and Applied Mathematics, and held at Potsdam, New York, 1972.
Includes bibliographies.
1. Wave motion, Theory of–Congresses.
2. Nonlinear theories–Congresses. 3. Differential equations, Nonlinear–Congresses. I. Newell, Alan C., 1941– ed. II. American Mathematical Society.
III. Society for Industrial and Applied Mathematics.
IV. Series: Lectures in applied mathematics (Providence) v. 15.
QA927.N66 515′.353 73–19504
ISBN 0–8218–1115–0

AMS (MOS) subject classifications. 34E20, 35-02, 35B10, 35G20, 35L60, 35L65, 35P25, 35R30, 35S10, 65M99, 73D99, 76B15, 76B25, 78A05.

Printed in the United States of America

Contents

* Speaker

Indexes

Foreword

The nineteen-sixties produced some major advances in the mathematical description of and our understanding of nonlinear wave propagation phenomena. Of particular significance were the methods of Whitham for describing the slow temporal and spatial modulation of fully nonlinear dispersive waves and of Gardner, Greene, Kruskal and Miura for finding the initial value solution to the Korteweg–de Vries equation. It was the purpose of this seminar to explore these and related theories and to exchange ideas about the most fruitful avenues of investigation for the immediate future.

The participants in the conference ranged over a wide spectrum of ages (from graduate students to senior scientists), background interests (biology, electrical engineering, geology, geophysics, mathematics, physics) and countries of origin (United States, Canada, Great Britain, Australia). A fairly heavy schedule was followed: organized lectures by the principal lecturers in the mornings and a full afternoon of informal and semi-organized sessions. A healthy and lively interaction between all participants prevailed for the full three and a half weeks. A reference library of relevant publications was extremely useful.

All participants deserve full marks for playing active roles in both the formal and informal meetings. A special thanks to Brooke Benjamin for rounding up volunteers for "Quaker-like" meetings which not only resulted in stimulating exchanges but also provided historical background and some anecdotes about the way in which some results were achieved. The good humor award must surely go to John Mahony from Western Australia who, after having had the misfortune to lose his luggage on route to Potsdam, made one (slightly yellowed) white shirt last for three weeks. Our thanks also to Bill Hall for whipping a motley group of very amateur athletes (including some ex-cricketers) into a respectable softball team.

Whereas we made every effort to invite all those who had made important contributions in the area of nonlinear wave propagation, we did overlook some of the recent work done by nonlinear optics groups in this country on the self-induced transparency (SIT) problem and some very useful investigations by colleagues in the U.S.S.R. and Japan. Discussion of these investigations arose during the conference and some references are included in the bibliography at the end of this volume. In addition, during the last year, there have been further significant advances and references to these studies are also given in the bibliography.

I want to express my deep appreciation to all those who helped support and organize this meeting: to the Mathematics Division of the National Science Foundation and the Office of Naval Research who provided financial support; to the American Mathematical Society and the Society for Industrial and Applied Mathematics who jointly sponsored the conference; to Gordon Walker, Lillian Casey and Carol Kohanski of the AMS for their great organizing efforts, to Margaret Reynolds who organized the Proceedings; to Clarkson College (particularly Chancellor Jack Graham and Dean Milton Kerker) for hosting the conference.

I would especially like to thank the Organizing Committee consisting of Mark Ablowitz, Victor Barcilon, Norman Bleistein, David Hector, Harvey Segur and Calvin Wilcox all of whom played major roles in making the seminar so enjoyable. Finally, those of us who participated in the meeting and who benefit from this volume are much indebted to the principal lecturers: Brooke Benjamin, David Benney, Martin Kruskal, Peter Lax, and Gerry Whitham.

ALAN C. NEWELL

Principal Lecturers

Lectures in Applied Mathematics
Volume 15, 1974

Lectures on Nonlinear Wave Motion

T. Brooke Benjamin

1. Introduction

These lectures will deal with a miscellaneous group of topics in the theory of nonlinear wave motion, but two connecting themes will run through the discussion. The first concerns the major ingredient of this subject describable as the *craft* of mathematical model-making. The aim is to highlight that the choice of approximate models for physical processes is often arbitrary, to be guided by considerations of expedience, common sense and style rather than by logical necessity. As the second theme, however, the role of *exact* theoretical results will be emphasized. The aim is to recommend the viewpoint that, once model equations are adopted, the rules of the game change essentially and conclusions drawn from these equations need to be supported as far as possible by mathematical rigour.

The range of problems covered in this article may be represented by the abstract evolution equation

$$u_t = Au, \tag{1.1}$$

with particular classifications of the nonlinear operator A. Here $u = u(x, t)$ stands for a function (or a matrix of functions) of position x in the physical domain and of time t, while Au stands for some nonlinear transformation of u qua function of x. (That is, we are concerned primarily with wave motion in systems whose basic properties do not change with time—the counterpart to "autonomous" systems with finite degrees of freedom, for

AMS (MOS) subject classifications (1970). Primary 76B25; Secondary 46N05.

which, in a representation like (1.1), A is just a function independent of time.) The equation thus determines a field of trajectories in an infinite-dimensional cylinder $D(A) \times [0, T]$, say, where $D(A)$ is the domain of definition of A in some appropriate function space and $[0, T]$ is the time-interval in question—which may be $[0, \infty)$ if the model of the evolutionary process remains meaningful over indefinitely large times. An initial-value problem amounts to the determination of the trajectory starting from a given point $u(x, 0)$ on the base of the cylinder.

Now, powerful general theories are available for evolution equations in which A belongs to certain classifications ("monotone," "accretive," "dissipative," etc.) that are typically representative of *diffusive* physical processes. Roughly speaking, the nonlinear equations in view are more akin to parabolic linear equations than to hyperbolic ones.[1] With regard to the present subject, however, this corpus of abstract theory is mostly unhelpful. We have to explain wave phenomena on which the effects of friction and other diffusive factors may appear to be secondary, so that it is reasonable to ignore such factors in developing theoretical models. But neglect of the factors that in nature ultimately ensure the smoothness of dependent variables may, of course, give to an idealized model some mathematically awkward properties. By use of common sense these may perhaps be interpretable as simulations of natural properties (e.g. the formation of shocks), but they may be incidentally troublesome features of the model. From even the simplest examples of frictionless models for nonlinear wave phenomena it can be seen that theoretical difficulties concerning the smoothness of solutions are typical of this subject, and so the lack of any comprehensive abstract theory is hardly surprising. Conversely, there is a general need for the careful mathematical appraisal of ad hoc models.

The present range of problems seems to be characterized most distinctly by the existence of conservation laws associated with the fundamental equation (1.1). In particular, there exist nonlinear scalar invariants for (1.1) (i.e. scalar functions of position in $D(A)$ that are constant along a solution trajectory). These typically correspond to such conserved quantities as total energy or momentum in the physical system. It can accordingly be said that the models in question are generally Hamiltonian systems, but this interpretation will not be given much attention in this article. In other respects, however, considerable use will be made of variational concepts.

In § 2 several aspects of the theory of water waves will be reviewed,

[1] For a review of modern research on the subject, reference may be made to Chapters 3–5 of the book by Carroll [1969].

particularly to illustrate the significance of steady wave motions and the derivation of approximate model equations. In § 3 some general issues entailed in the appraisal of approximate models will be discussed, with particular reference to Hadamard's criteria applied to equations of the Korteweg-de Vries type. In § 4 the stability of steady waves will be considered. Finally, in § 5, there will be a discussion of the rather tricky problem of establishing steady-wave solutions of certain generalized evolution equations introduced in § 3.

2. The water-wave problem

This problem has been pivotal in the history of nonlinear wave theory. The archetypal model equations of Korteweg and de Vries and of Boussinesq, for example, were originally derived as approximations for water waves, and research into the problem has been sustained vigorously up to the present day. The problem is recalled now in order to bring out several points of general interest. We consider it in its simplest version, that referring to two-dimensional flow of a perfect liquid over a horizontal bottom, but some generalizations will be mentioned later.

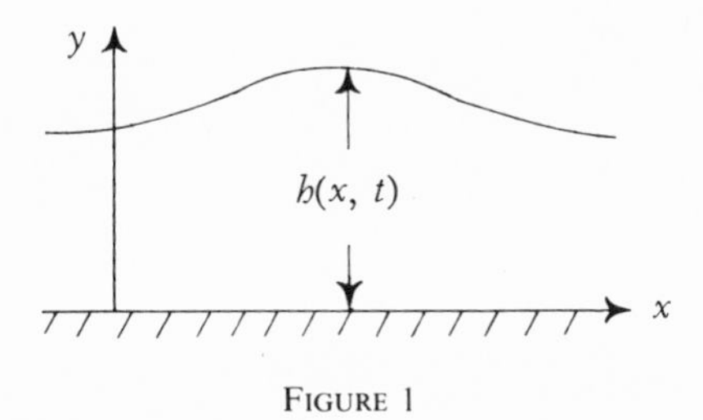

FIGURE 1

In terms of axes (x, y) as defined in Figure 1, the dependent variables are the velocity potential $\phi(x, y, t)$ and the height $h(x, t)$ of the free surface above the bottom $y = 0$. The density of the liquid is taken to be unity. The specifications of the problem are then

$$\phi_{xx} + \phi_{yy} = 0 \quad \text{in } 0 \leqq y \leqq h(x, t), \ -\infty < x < \infty, \tag{2.1}$$

$$\phi_y = 0 \quad \text{on } y = 0, \tag{2.2}$$

$$h_t + h_x\phi_x - \phi_y = 0 \quad \text{on } y = h(x, t), \tag{2.3}$$

$$\phi_t + \tfrac{1}{2}(\phi_x^2 + \phi_y^2) + gh - T\sigma = \text{const.} \quad \text{on } y = h(x, t), \tag{2.4}$$

together with initial conditions and asymptotic conditions for $x \to \pm\infty$ when appropriate (cf. Stoker [1957, § 1.4]). Here T is the coefficient of surface tension, and $\sigma = h_{xx}/(1 + h_x^2)^{3/2}$ is the curvature of the free surface.

2.1. Exact results. Although a great deal of useful information has been accumulated in approximate treatments of this problem and elaborations of it, very few exact results are available. They include the following, for the case $T = 0$:

(1) Krasovskii [1961] has shown that, for every value of max $|h_x|$ in the open interval $(0, 1/\sqrt{3})$, a periodic solution exists representing a steady progressive wave-train. This result supercedes several earlier, constructive proofs of existence (e.g. by Levi–Civita [1925] in the case of infinite depth) for periodic waves of sufficiently small amplitude. Krasovskii's proof is based on a topological theorem concerning cones in Banach spaces, and it must be counted as one of the most specular accomplishments of functional analysis applied to hydrodynamical problems. It is now known, however, at least in the case when max $|h_x|$ is small and the ratio of wavelength to mean depth is not too large, that the periodic solutions in question represent unstable states of motion (Benjamin and Feir [1967]; Benjamin [1967b]).

(2) Friedrichs and Hyers [1954] have proved the existence of solitary-wave solutions with small amplitude. In their treatment, as in constructive analyses of periodic waves, the restriction to small amplitude is necessary to establish convergence of a scheme of successive approximations—the first stage of which recovers approximate results found by Boussinesq and by Rayleigh in the 1870's. According to experimental and computational evidence, it is very likely that permanent solitary waves exist throughout the same range of max $|h_x|$ as has been established for periodic waves, but this has yet to be proved. [It may be of interest to note the technical reason why the method used in Krasovskii's global theory of periodic waves is unavailing for solitary waves. In his treatment the boundary-value problem is recast as a nonlinear operator equation (often called the Nekrasov equation). A positive cone in a suitable Banach space (of functions defined over one wave period) is shown to be mapped into itself by the operator, and this mapping is shown to be completely continuous (compact); hence the topological theorem mentioned above is confirmed to be applicable. For the solitary-wave problem, a corresponding operator equation holds and similarly an invariant cone can be found in a Banach space of functions defined on the respective *infinite* domain. But the operator is not a completely continuous mapping of this cone, and for just this reason the crucial theorem is no longer applicable.]

(3) Several exact results have been established concerning waves excited by a steady forcing factor—for example, periodic waves caused

by sinusoidal displacements of the lower boundary, and small-amplitude solitary waves driven by submerged vortices. For a review, reference may be made to the article by Wehausen [1965, pp. 543–548].

(4) Various qualitative conclusions, particularly about questions of nonexistence, can be established by simple arguments of the type common in "open-channel hydraulics." The literature on the subject is dismayingly diffuse, and it must suffice here to mention just one result. In the perfect-fluid model, no *steady* free wave exists in the form of a transition between different asymptotic levels of the free surface as illustrated in Figure 2.

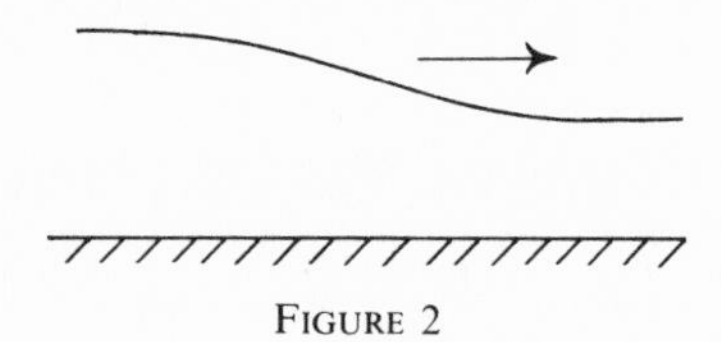

FIGURE 2

Equations (2.1)–(2.4) imply the conservation of mass, energy and horizontal momentum, and it is easily verifiable that all three conditions cannot be satisfied by such a steady wave motion. This result is analogous to the well-known principle in gas dynamics, that no steady wave-front exists conserving mass, momentum and available energy (i.e. in a shock wave the latter necessarily decreases). Since steady periodic waves are possible in the present system, the analogy with gas dynamics is evidently limited, and on the strength of experimental evidence one might plausibly conjecture the existence of steady undular shocks, like Figure 2 on the right but asymptotic to a periodic wave-train on the left. It is known, however, that a steady wave motion of this form is also impossible in the absence of energy dissipation (see Benjamin and Lighthill [1954]).

In respect of the initial-value problem for the system (2.1)–(2.4), virtually nothing in the way of rigorous theory is available. Moreover, some degree of mathematical intractability seems inevitable here. We recognize the probability that the general initial-value problem cannot be correctly posed (well set), because we know that in practice water waves may *break*—that is, the motion may become turbulent and so lose continuous dependence on initial data. This aspect of the subject still remains largely mysterious, and reservations regarding it are needed to put any theoretical work on water waves into a properly scientific perspective. The fact that most existing theory—e.g. linearized and long-wave approximations—is essentially tentative does not, of course, impair its practical value. Satisfactory accounts of many real phenomena are provided, even though the accuracy of approximations generally cannot be proved a priori. In the final resort one uses experimental checks to confirm reliability.

2.2. A variational formulation. Interesting uses for variational methods in the water-wave problem have been found by Whitham [1967a], [1967b], Luke [1967] and others. These generally entail Hamilton's principle or other variational principles of the same kind where the varied functional is the integral of a suitable "Lagrangian function" over the product of the flow domain and a time interval. We now consider a quite different application of variational methods which reveals several new aspects of the general problem, particularly regarding the significance of permanent waves.

For a motion that vanishes sufficiently rapidly as $x \to \pm\infty$, at each of which limits the depth of the stationary liquid is H and ϕ is a constant, the total energy is defined by

$$E = \int_{-\infty}^{\infty}\int_{0}^{h} \tfrac{1}{2}(\phi_x^2 + \phi_y^2)\,dy\,dx + \int_{-\infty}^{\infty} \tfrac{1}{2}g(h - H)^2\,dx + \int_{-\infty}^{\infty} T\{(1 + h_x^2)^{1/2} - 1\}\,dx. \tag{2.5}$$

The first integral expresses the kinetic energy, the second the potential energy entailed in displacing the free surface from the level $y = H$ of equilibrium under gravity, and the third integral expresses the energy associated with the total extension of the free surface. Correspondingly, the total horizontal momentum (impulse) of the motion is defined by

$$\begin{aligned} I &= \int_{-\infty}^{\infty}\int_{0}^{h} \phi_x\,dy\,dx = H[\phi]_{-\infty}^{\infty} - \int_{-\infty}^{\infty} \phi_S h_x\,dx \\ &= H[\phi]_{-\infty}^{\infty} + I'\ (\text{say}) = \int_{-\infty}^{\infty} (\phi_S)_x h\,dx. \end{aligned} \tag{2.6}$$

(Here we introduce the notation $\phi_S(x, t) = \phi(x, h, t)$ for the value of ϕ at the free surface.) An important fact is that E, I and I' are *invariants* of the motion, as may be confirmed from equations (2.1)–(2.4) (cf. Lamb [1932, § 10]).

We now express the first variation of E and I', noting that

$$\delta\phi_S = \delta\phi(x, h, t) = (\delta\phi)_S + (\phi_y)_S\delta h.$$

The variation δh is supposed to vanish as $x \to \pm\infty$, but $[\delta\phi]_{-\infty}^{\infty}$ need not vanish.[2] Using Green's theorem together with (2.1) and 2.2), we obtain

[2] Note that if we impose the extra, artificial constraint that the latter quantity be zero, the variational principle can be expressed in terms of I rather than I'. I am indebted to Professor M. S. Longuet-Higgins for some helpful comments about the significance of the constituent of I' of impulse for two-dimensional wave motions.

$$\delta E = \int_{-\infty}^{\infty} \left[(\phi_y - \phi_x h_x)_S \delta\phi_S + \left\{ \frac{1}{2}(\phi_x^2 + \phi_y^2)_S + g(h - H) - \frac{T h_{xx}}{(1 + h_x^2)^{3/2}} - [(\phi_y - \phi_x h_x)\phi_y]_S \right\} \delta h \right] dx, \tag{2.7}$$

$$\delta I' = \int_{-\infty}^{\infty} \{ -h_x \delta\phi_S + (\phi_S)_x \delta h \} \, dx. \tag{2.8}$$

If the pair of functions $\phi_S(x, t)$ and $h(x, t)$ are considered as a matrix (ϕ_S, h), (2.7) and (2.8) can be represented in the form

$$\delta E = \int_{-\infty}^{\infty} (\text{grad } E)\delta(\phi_S, h) \, dx, \tag{2.7$'$}$$

$$\delta I' = \int_{-\infty}^{\infty} (\text{grad } I')\delta(\phi_S, h) \, dx \tag{2.8$'$}$$

(which defines the meaning of 'grad' adopted here). Hence the nonlinear boundary conditions (2.3) and (2.4) are seen to be equivalent to

$$\partial_t(\text{grad } I') + \partial_x(\text{grad } E) = 0. \tag{2.9}$$

[Note that the first component of (2.9) is just the first derivative of (2.3) with respect to x. Since $h \to H$ as $x \to \infty$, (2.3) is therefore implied. To check the second component, observe that $\text{grad}_h E = \frac{1}{2}(\phi_x^2 + \phi_y^2)_S + g(h - H) - T\sigma - [(\phi_y - \phi_x h_x)\phi_x]_S = -[\phi_t + \phi_y h_t]_S = -(\phi_S)_t$, $\text{grad}_h I' = (\phi_S)_x$.]

The result (2.9) is interesting in a variety of ways as follows:

(1) Coupled with the boundary condition (2.1) on the bottom and the asymptotic conditions for $x \to \pm\infty$, the functions $\phi_S(x, t)$ and $h(x, t)$ determine the motion completely; for at each t the velocity potential ϕ is determined as the unique solution of a linear boundary-value problem. It is therefore legitimate to consider E as well as I' to be a functional of ϕ_S and h (qua functions of x). Thus (2.9) implicity represents a pair of equations for ϕ_S and h, the y-dependence of the original problem having been eliminated. Since these are equations of evolution to which initial conditions can be directly referred, this idea may have some potentiality for an exact treatment of the initial-value problem. As noted earlier, however, the possibility of wave breaking gravely complicates the issue, and the question of what restrictions will be necessary for a mathematically adequate theory remains very obscure.

(2) The various model equations to be considered presently all have the form of (2.9)!

(3) *Modification for periodic motions.* In the case of wave motions that

are periodic in x, the definitions (2.5) and (2.6) are meaningless, but a result corresponding to (2.9) is obtained when E and I' are redefined by integrals over one period. The periodicity condition replaces the condition at infinity in the steps leading to (2.9), but in all other respects the derivation is the same.

(4) *Generalization to three dimensions.* The present formulation can easily be extended to the problem of three-dimensional wave motions, where in terms of Cartesian axes (x_1, x_2, y) the bottom is the horizontal plane $y = 0$ and the free surface is $y = h(x_1, x_2, t)$. The total energy E is again a scalar, of course, but the horizontal momentum is a vector with components I_i parallel to x_i. Using the notation

$$\delta E = \int_{-\infty}^{\infty} \int_{-\infty}^{\infty} (\text{grad } E)\delta(\phi_S, h)\, dx_1\, dx_2,$$

where as before (grad E) stands for a two-component matrix, one finds that the hydrodynamical problem is representable by

$$\partial_t(\text{grad } I_i) + \partial_{x_i}(\text{grad } E) = 0.$$

(5) *Case of an uneven bottom.* When the bottom is not level, being described by the equation $y = b(x)$, the boundary condition (2.2) is replaced by $\partial_n \phi = 0$ on $y = b(x)$, and in the definitions (2.5) and (2.6) of E and I' the lower limit of integration with respect to y is replaced by $b(x)$. It is found in this case that (2.9) still applies, although the implicit dependence on ϕ_S is not the same as before. Also, while E is still a constant, the corresponding I' is no longer a constant (since the hydrodynamic pressure on the bottom can produce a net horizontal force).

(6) It is noteworthy incidentally that $I = \int h\bar{u}\, dx$, where $\bar{u}$ is the horizontal mean velocity. We also have that $I = \int h(\phi_S)_x\, dx$; but $\bar{u}$ is *not* identical with $(\phi_S)_x$.

(7) *The most interesting aspect.* For *steady* wave motions we have $\partial_t \equiv -C\partial_x$, where C is the velocity of translation in the $+x$ direction. So in this case (2.9) reduces to

$$\text{grad } E = C \text{ grad } I', \tag{2.10}$$

which is the Euler–Lagrange equation for the conditional variational problem $\delta E = 0$ for I' fixed, or for the dual problem $\delta I' = 0$ for E fixed. The velocity C thus takes the role of a Lagrangian multiplier.

For the physical model represented by (2.1)–(2.4), the only steady motions evanescent at infinity appear to be the class of solitary waves. Presumably, therefore, a solitary wave realizes the *minimum* total energy corresponding to a given I' (or maximum I' for a given E). Similarly, among possible periodic motions with a given wavelength λ, steady periodic waves presumably realize the minimum of E_λ (total energy per wavelength) for a given I'_λ.

The present aspects of the water-wave problem are indicative of a general interpretation that will be made precise later, with reference to a simpler model of conservative nonlinear wave motion. The evolutionary process may be regarded as a trajectory, determined by (2.9) on the intersection of two hypersurfaces (manifolds) $E = \text{const.}$ and $I' = \text{const.}$ in some infinite-dimensional space, each point of which represents a pair of functions $\phi_S(x)$ and $\zeta(x) = h(x) - H$, and the zero point of which corresponds to the undisturbed state of the physical system. The vectors grad E and grad I' are normal to the respective surfaces, and are collinear only in the case of steady translatory motions.

2.3. Implications concerning stability. We note here some plausible conjectures linked with the extremal property of steady waves. As yet, the original water-wave problem has been found too complicated for these propositions to be proved, but it will be shown later how they are confirmed for simpler models simulating long water waves.

Let $u(x, t)$ stand for a pair of functions $\phi_S(x, t)$, $\zeta(x, t) = h - H$ that comprise a solution of (2.9). As noted above, it may reasonably be supposed that a solitary wave, say $u = \psi(x - Ct)$, realizes a minimum of $E(u)$ for a given $I'(u)$. To express this property more definitely, a metric $d(v, w)$ needs to be introduced, measuring distance between any two functions v, w of x in the class to which $u(x, t)$ belongs at every t. Considering the totality of functions for which $I'(u) = I'(\psi)$, and supposing that functions that are just translations of ψ are excluded from competition for the minimum of E,[3] we may then assert that in some neighbourhood of ψ, say $d(u, \psi) \leqq \varepsilon$, there holds an inequality in the form

$$E(u) - E(\psi) \geqq f\{d(u, \psi)\}, \tag{2.11}$$

where $f(d)$ is a monotonic nondecreasing function with $f(0) = 0$ and $f(\varepsilon) > 0$.

It may also be supposed that the difference $E(u) - E(\psi)$ can be made arbitrarily small by choosing initial values $u_0 = u(x, 0)$ close enough in some sense to ψ_0. Specifically, there is some suitable metric D such that $E(u_0) - E(\psi_0) \to 0$ as $D(u_0, \psi_0) \to 0$.

Since E and I' are invariants of any solution u, it follows that a solitary wave ψ is *stable* according to the Liapunov definition (see § 4). More precisely, the preceding conclusions imply that ψ is stable, with respect to the considered metrics, for perturbations satisfying $I'(u) = I'(\psi)$. To

[3]An account of how this can be done—in fact by a suitable choice of metric—will be given in § 4.

show that a solitary wave is stable for all perturbations, one needs merely to confirm that in a d-neighbourhood of ψ, however small, the values of $I'(u)$ differing from $I'(\psi)$ are realizable by other solitary waves which can be made arbitrarily close to ψ by restricting them to a D-neighbourhood of ψ_0. Thus every perturbation with $I'(u) \neq I'(\psi)$ can be considered as a perturbation that has the same I as a neighbouring solitary wave, and the foregoing stability result can be applied generally. [It is noteworthy that these ideas are comparable with ideas developed by Arnol'd [1963], [1965] concerning the stability of inviscid rotational flows with fixed boundaries.]

In the case of wave motions that are periodic in x, similar considerations apply with regard to the invariants E_λ and I'_λ. Let ψ_λ denote the steady periodic wave-train whose fundamental period is λ. Since $E_\lambda(\psi_\lambda)$ is a minimum among the class of functions with the same period and with $I'_\lambda(u) = I'_\lambda(\psi_\lambda)$, it may be concluded that ψ_λ is stable for perturbations in this class, and stability for general perturbations with period λ may also be inferred by the argument indicated above. But the minimum property cannot be expected to hold among a class of functions with sufficiently *longer* period.[4] The question thus arises whether, when released for the present sufficient condition for stability (i.e. restriction to perturbations with the same period), periodic steady waves become *un*stable. The answer appears to be that periodic water waves are always unstable in some way, and the same is probably true in general for other dispersive nonlinear systems.

2.4. Approximate models. A vast amount of theoretical work has been done recently on the approximate evolution equations, applying to long water waves of small amplitude, that were originally proposed by Boussinesq and Korteweg and de Vries (KdV). In the context of water waves as in the many other contexts where they have been found to apply, these equations have been derived in various ad hoc ways, and although none of the available derivations is rigorous some fairly careful accounts of the issues have appeared (e.g. Meyer [1967]). To do something mildly original in covering this well-trodden ground, the following derivation is based on the preceding variational formulation (cf. Whitham [1967b, § 2]).

For simplicity of illustration, the effects of surface tension are now ignored. Putting $V = (\phi_S)_x$, we have exactly from (2.6) that

$$\text{grad}_h I = V, \qquad \text{grad}_V I = h. \tag{2.12}$$

[4] Because, for instance, Jacobi's necessary condition for a conditional minimum will not hold over a sufficiently extended domain. For a proof of the above claim in respect of a simpler model, see § 4.5.

The whole burden of the problem thus lies in estimating the kinetic-energy component of E, i.e.,

$$K = \int_{-\infty}^{\infty} \int_0^h \tfrac{1}{2}(\phi_x^2 + \phi_y^2)\, dy\, dx. \tag{2.13}$$

To formalize the scheme of approximation, scaled variables should be introduced. But this would necessitate a lot more writing while scarcely adding to the essential sense of what is done. The underlying assumption is that

$$(h - H)/H = O(\varepsilon), \qquad h_x, h_t/(gH)^{1/2} = O(\mu\varepsilon),$$

where ε and μ, the inverse of the characteristic horizontal scale for the motion, are both $\ll 1$. The operations ∂_x, ∂_t carry the implication $O(\mu)$, and we shall finally take $O(\varepsilon)$ and $O(\mu^2)$ to be equivalent.

The function $\phi(x, y, t)$ is harmonic in (x, y) and satisfies $\phi_y(x, 0, t) = 0$. Accordingly, we can introduce a function $f(x, t)$ such that

$$\underset{}{\phi_x} = \underset{(\varepsilon)}{f} - \underset{(\varepsilon\mu^2)}{\tfrac{1}{2}y^2 f_{xx}} + O(\varepsilon\mu^4), \qquad \phi_y = \underset{(\varepsilon\mu)}{-y f_x} + O(\varepsilon\mu^3).$$

Hence

$$\phi_x^2 + \phi_y^2 = f^2 + y^2(f_x^2 - f f_{xx}) + O(\varepsilon^2\mu^4),$$

and from (2.13), after an integration by parts,

$$K = \int_{-\infty}^{\infty} \{\tfrac{1}{2} f^2 h + \tfrac{1}{3} h^3 f_x^2 + O(\varepsilon^2\mu^4)\}\, dx.$$

Since

$$V = (\phi_x)_S + (\phi_y)_S h_x = f - \tfrac{1}{2}h^2 f_{xx} + O(\varepsilon^2\mu^2, \varepsilon\mu^4),$$

we have, to the same order of approximation,

$$f = V + \tfrac{1}{2}h^2 V_{xx}.$$

Hence we obtain finally

$$E = \int_{-\infty}^{\infty} \{\tfrac{1}{2}V^2 h - \tfrac{1}{6}h^3 V_x^2 + \tfrac{1}{2}g(h - H)^2 + O(\varepsilon^2\mu^4, \varepsilon^3\mu^2)\}\, dx.$$

When this approximation is used together with (2.12) in evaluating (2.9), the result is

$$\delta V: \qquad h_t + (Vh)_x + \tfrac{1}{3}(h^3 V_x)_{xx} = 0,$$

$$\delta h: \qquad V_t + V V_x + g h_x = 0,$$

where the error in both equations is $O(\varepsilon^2\mu^3, \varepsilon\mu^5)$. Introducing the mean velocity

$$\bar{u} = \frac{1}{h}\int_0^h \phi_x \, dy = f - \tfrac{1}{6}h^2 f_{xx} + \cdots = V + \tfrac{1}{3}h^2 V_{xx} + \cdots,$$

in terms of which therefore $V = \bar{u} - \frac{1}{3}h^2\bar{u}_{xx} + \cdots$, we hence obtain, to the same order of approximation,

$$\text{(2.14)} \qquad \begin{aligned} h_t + (\bar{u}h)_x &= 0, \\ \bar{u}_t + \bar{u}\bar{u}_x + gh_x - \tfrac{1}{3}H^2\bar{u}_{xxt} &= 0. \end{aligned}$$

These are the Boussinesq equations (cf. Keulegan and Patterson [1940]). Note that the first of (2.14) recovers the exact condition of volume conservation in the liquid, and the fractional error in the second equation is of second order in ε and μ^2.

To remove physical factors that are henceforth unessential, take H as the unit of length and $(H/g)^{1/2}$ as the unit of time. Also let us write $h = 1 + \zeta$, and just u ($\equiv \bar{u}/(gH)^{1/2}$) for the second dependent variable in (2.14). The equations are then as follows, where the orders of magnitudes of the terms are indicated:

$$\text{(2.14')} \qquad \underset{(\varepsilon\mu)}{\zeta_t} + \underset{(\varepsilon\mu)}{u_x} + \underset{(\varepsilon^2\mu)}{(u\zeta)_x} = 0, \qquad \underset{(\varepsilon\mu)}{u_t} + \underset{(\varepsilon\mu)}{\zeta_x} + \underset{(\varepsilon^2\mu)}{uu_x} - \underset{(\varepsilon\mu^3)}{\tfrac{1}{3}u_{xxt}} = 0.$$

To obtain a single equation governing one-way propagation in the $+x$ direction, we first note that the rudimentary approximation for small ε and μ^2 is $\zeta_t + u_x = 0$, $u_t + \zeta_x = 0$. The general solution of this simple system representing positively travelling waves is $\zeta = u = F(x - t)$, where F is any differentiable function, and thus $\partial_t \equiv -\partial_x$ to a first approximation.[5] Hence for the case of unidirectional propagation it may be seen that, to the order of approximation implicit in (2.14′), these equations are formally equivalent to $\zeta = u + \frac{1}{4}u^2 - \frac{1}{6}u_{xx}$ and

$$\text{(2.15)} \qquad u_t + (1 + \tfrac{3}{2}u)u_x - \tfrac{1}{6}u_{xxt} = 0$$

or

$$\text{(2.16)} \qquad u_t + (1 + \tfrac{3}{2}u)u_x + \tfrac{1}{6}u_{xxx} = 0.$$

[5] Note that this rudimentary estimate is essential to *all* derivations of the KdV equation or equivalent model equations, in whatever physical context. That is, the assumption of unidirectional propagation underlies all physical applications of the KdV equation.

The latter is the KdV equation, but (2.15) has exactly the same formal status as an approximation for small ε and μ^2. (Indeed, yet other formally equivalent equations could be written down.) Equation (2.15) has been studied in a recent paper by Benjamin, Bona and Mahony [1972], and for convenience it will henceforth be referred to as the BBM equation. Note that the coefficients $\frac{3}{2}$ and $\frac{1}{6}$ in (2.15) and (2.16) can be absorbed into redefinitions of the dependent and independent variables, so we lose no generality by taking the tidied forms of these equations in the subsequent discussion.

In the original derivation by Korteweg and de Vries [1895], the effects of surface tension were included. The main difference due to surface tension is that the coefficient of the third derivative in (2.15) or (2.16) is multiplied by a factor $1 - (3T/gH^2)$, which becomes negative when H is small enough (less than about 4.6 mm for water). The consequences of this change of sign have interesting physical interpretations, but the case appears exceptional among the various applications of the KdV equation.

Other physical systems to which the KdV equation applies as a long-wave approximation—and to which accordingly the BBM equation equally well applies—include the following:

1. Internal waves in a density-stratified fluid bounded by horizontal planes.
2. Acoustic-gravity waves in a compressible heavy fluid.
3. Axisymmetric waves in a nonuniformly rotating fluid with cylindrical boundary.
4. Planetary waves in β-plane models.
5. Plane compressional waves in bubbly liquids.
6. Axisymmetric waves in rubber cords.
7. Hydromagnetic waves in cold plasmas.
8. Ion-acoustic waves.
9. Acoustic waves in anharmonic crystals.

3. Appraisal of approximate models

As exemplified in § 2.4, schemes of approximation for simplifying nonlinear wave problems usually have as end-product a model equation that can be expressed in several alternative but formally equivalent forms. In the case of solutions that happen to satisfy the assumptions on which the approximate model is derived, the alternatives are virtually the same. But in general the different forms of the model have different mathematical properties, some of which may be specially interesting—or troublesome—incidentally to the model's primary purpose. Although various criteria may reasonably be adopted to discriminate between the alternatives, it is clear there can be no absolute answer to questions of preferability. At least from the standpoint of expedient research into the original physical problem, however, there is always a good case for preferring a model equation that is amenable to, first, a straightforward qualitative theory and, second, easy computation of solutions.

To assess the basic mathematical fitness of a model equation of evolution, the chief implement is Hadamard's famous definition of the requirement that an initial-value problem be *well set*. In order to express this neatly, the process of solving the problem may be described as an evolution operator $\boldsymbol{U}$ that takes elements in a suitable class of initial data into a class of solutions. Then the requirement is that, for specified function classes, $\boldsymbol{U}$ be defined as a unique and continuous transformation.

To illustrate the Hadamard concept, let us consider a result given in the paper by Benjamin, Bona and Mahony [1972] concerning the problem

$$\begin{aligned} u_t + u_x + uu_x - u_{xxt} &= f(x, t), \\ u(x, 0) &= g(x) \qquad (x \in \boldsymbol{R}). \end{aligned} \tag{3.1}$$

Here the BBM equation is modified by the prescribed term $f(x, t)$, which could represent some kind of forcing action on the physical system in question. Some notation for function classes now needs to be introduced.

Let $\mathscr{C}_T$ $[\equiv C\boldsymbol{R} \times ([0, T])]$ denote the class of functions $u(x, t)$ that are continuous and uniformly bounded on the infinite strip $\boldsymbol{R} \times [0, T]$, and let $\mathscr{C}_T^{l,m}$ denote the narrower class of functions such that $\partial_x^i \partial_t^j u \in \mathscr{C}_T$ for $0 \leqq i \leqq l, 0 \leqq j \leqq m$. These function classes become Banach spaces under the usual supremum norms. We further write $W_2^1(\boldsymbol{R})$ for the Sobolev space of square-integrable (L_2) functions with generalized first derivatives that are also square-integrable [see § 4.2 below for the definition of the norm: this space is also commonly denoted by $H^1(\boldsymbol{R})$]. Next, $\mathscr{L}_T$ is written for the class of functions $f(x, t)$ defined on $\boldsymbol{R} \times [0, T]$ such that $f \in L_2(\boldsymbol{R})$ for each $t \in [0, T]$, and such that $[0, T]$ is mapped continuously into $L_2(\boldsymbol{R})$. This becomes a Banach space under a norm defined as the supremum of the L_2 norm for $0 \leqq t \leqq T$ (cf. BBM [1972, p. 62]). Similarly, a Banach space $\mathscr{W}_T$ may be defined whose elements are functions $u(x, t)$ taking $[0, T]$ continuously into $W_2^1(\boldsymbol{R})$.

We suppose that the prescribed conditions of the problem (3.1) are classified as follows. First, the initial function $g(x)$ belongs to both $C^2(\boldsymbol{R})$ and $W_2^1(\boldsymbol{R})$. Second, the forcing function $f(x, t)$ belongs to both $\mathscr{C}_T$ and $\mathscr{L}_T$, for some arbitrarily large but finite $T > 0$. Accordingly, the prescribed conditions may be considered to belong to the topological product

$$\mathscr{G} = C^2(\boldsymbol{R}) \cap W_2^1(\boldsymbol{R}) \times \mathscr{C}_T \cap \mathscr{L}_T,$$

each element of which is a pair of functions (g,f), where $g(x) \in C^2(\boldsymbol{R}) \cap W_2^1(\boldsymbol{R})$ and $f(x,t) \in \mathscr{C}_T \cap \mathscr{L}_T$. $\mathscr{G}$ becomes a Banach space by defining its norm to be

$$\|(g,f)\|_{\mathscr{G}} = \|g\|_{C^2 \cap W_2^1} + \|f\|_{\mathscr{C} \cap \mathscr{L}}.$$

The first term on the right-hand side may be defined as the sum of the C^2 and W_2^1 norms, or as the greater of the two, and the second term may be defined similarly [BBM, p. 70].

Now, the result in question establishes that, for any $(g,f) \in \mathscr{G}$, there is a unique solution $u(x,t)$ of (3.1) which belongs to both $\mathscr{W}_T$ and $\mathscr{C}_T^{2,m}(\boldsymbol{R})$, where m can be assigned any value $\geqq 1$ (see BBM, footnote on p. 63). That is, the class of solutions is $\mathscr{S} = \mathscr{C}_T^{2,m} \cap \mathscr{W}_T$, which too can be defined as a Banach space if m is given any finite value. The complete result may be stated: *A mapping* $U: \mathscr{G} \to \mathscr{S}$ *is defined assigning to each* $(g,f) \in \mathscr{G}$ *a unique solution of* (3.1), *and this mapping is continuous*. By *continuous* we mean here that for any particular element (g_0, f_0) and any number $\varepsilon > 0$, however small, a number $\delta > 0$ can be found such that

$$\|(g_0, f_0) - (g,f)\|_{\mathscr{G}} \leqq \delta$$

implies $\|u_0 - u\|_{\mathscr{S}} \leqq \varepsilon$. (Some essentials of the BBM theory will be outlined in § 3.3 below.)

Notwithstanding the complicated symbolism needed to express these properties of the BBM model, the general idea behind them is quite simple, and indeed properties of this kind (assumed even if not proved!) appear prerequisite to the *usefulness* of any model equation for real wave phenomena. It is obviously necessary that solutions exist in some appropriate classification, corresponding to reasonable classes of prescribed data, also that each solution be unique in order to be meaningful as a physical approximation. And it is also a natural requirement that solutions depend continuously on the prescribed conditions, as do the real phenomena that are simulated. It is possible, of course, that a model equation may cease to have any kind of validity after some finite time-interval; for waves may "break" in some way, at which time the Hadamard criteria may no longer be satisfied (at least not without additional specifications for the model). But, to be useful at all, a model problem must be well set in some clearly definable, even if restricted sense. The point being made is that results of the kind illustrated above are not just mathematical niceties. If a model problem is not well set in an appropriate sense, then it is valueless in practical respects: for instance, attempts to compute solutions numerically will be futile.

There is another requirement that may reasonably be juxtaposed with the Hadamard criteria, and this is describable as "continuity of the modelling." In other words, small changes in the specifications of a model equation—not only in numerical parameters but also in the functional operations involved—should have small effects on solutions over a whole function class. So far we have only been concerned with differential equations, and variations among classes of differential operators do not admit a suitable notion of smallness. In fact, the difference between the KdV and BBM equations is gross with regard to their general solutions, and it is only for extremely long waves that they are comparable. However, small perturbations of either equation can be defined in the present sense by considering the respective operators as elements of classes of pseudo-differential operators, and this interpretation seems quite important with regard to the physical meaning of these equations. The matter will be explained in the next section.

3.1. Significance of long-wave models. As was noted in § 2.4, the KdV equation has been derived as a long-wave approximation in many different physical contexts, by various ad hoc methods. Certain common essentials can be recognized, however, and we shall now review these in order particularly to emphasize the arbitrary status of this and comparable models such as the BBM equation (cf. BBM, § 2 and Appendix 1).

If finite-amplitude effects are ignored, the dispersive properties of a uniform system can generally be calculated without further approximation. That is, a linearized theory of travelling waves can be developed without restriction on wavelength (as is well known for water waves, plasma waves, etc.). The primary result is a *dispersion relation*, $c = c(k)$, between the phase velocity c and wave-number $k\,(=2\pi/\text{wavelength})$ of sinusoidal waves. Hence, by appeal to Fourier's principle, it may be inferred that a general wave motion $u(x, t)$, defined for $-\infty < x < \infty$ and characterized by unidirectional propagation in the $+x$ direction, satisfies the equation

$$u_t + (\boldsymbol{L}u)_x = 0, \tag{3.2}$$

in which $\boldsymbol{L}$ is a linear operator defined as follows. Let $\hat{u}(k)$ denote the Fourier transform of $u(x)$, i.e. $\hat{u} = \mathscr{F}u$ and $u = \mathscr{F}^{-1}\hat{u}$. Then

$$(\boldsymbol{L}u)\hat{} = c(k)\hat{u}(k), \qquad \text{i.e. } \boldsymbol{L}u = \mathscr{F}^{-1}(c\mathscr{F}u). \tag{3.3}$$

For all the systems in question, the function $c(k)$ has a smooth maximum with nonzero curvature at $k = 0$, say $c(0) = 1$ and $c''(0) = -2\gamma^2$. Thus an approximation for sufficiently small k (i.e. sufficiently long waves) is

$$c(k) \doteqdot 1 - \gamma^2 k^2,$$

corresponding to which the definition (3.3) formally gives

$$\boldsymbol{L} \doteqdot \boldsymbol{I} + \gamma^2 \partial_x^2.$$

On the other hand, if frequency-dispersive effects are ignored, nonlinear effects on long waves propagating in the $+x$ direction appear to be such as to make a positive wave progressively steepen ahead of its crest, a form of behaviour represented by the equation $u_t + u_x + uu_x = 0$. Hence, supposing that mild nonlinear and mild dispersive effects are independent in a first-order approximation, one may combine the two estimates and at once infer the KdV equation

$$u_t + u_x + uu_x + \gamma^2 u_{xxx} = 0.$$

The least requirement for the validity of this approximation is, of course, that the third and fourth terms on the left-hand side be numerically small compared with the first two.

To obtain the BBM equation by the same argument, one may introduce $v = \boldsymbol{L}u \doteqdot u - \gamma^2 u_{xx}$, whereupon (3.2) becomes approximately

$$(v - \gamma^2 v_{xx})_t + v_x = 0.$$

The correction for small nonlinear effects then enters as before.

Other model equations can be written down with equal justification. One possibility suggested by Whitham [1967b] is to keep intact the exact dispersion operator $\boldsymbol{L}$ according to linearized theory, but to simulate nonlinear effects by the first-order approximation appropriate to long waves. Thus one has $u_t + (\boldsymbol{L}u)_x + uu_x = 0$. Since $c(k)$ is generally a bounded function such that $c(k) \to 0$ for $|k| \to \infty$, $\boldsymbol{L}$ is a smoother operator than identity, in fact obliterating small-scale features; but this equation allows no comparable moderation of nonlinear effects on such features. It is therefore to be expected that, for any initial wave form with significant short-wave components, a real solution will exist for only a short time.

This form of equation has been investigated by Seliger [1968] as a model for the breaking of waves. The concept is interesting because, for example, water waves with large slopes can certainly break, so that eventual failure may be a quite legitimate feature of a perfect-fluid model. The conclusions drawn from such equations are dubious, however, in that there is little hope of them correctly simulating the nonlinear effects in the natural breaking process. No practical evidence exists that moderately long waves of moderately small amplitude break, at least not when the local length and amplitude scales are such as to give any justification for approximations of the present kind; and so it seems preferable on the whole to adopt models free from spurious breaking phenomena. In this respect a more satisfactory model incorporating the exact dispersion operator $\boldsymbol{L}$ would be

$$u_t + \boldsymbol{L}(u_x + uu_x) = 0,$$

which can be shown to have comparatively well-behaved solutions. This has the same formal status as the preceding equations, because to a first approximation for long waves $\boldsymbol{L}$ is the same as the identity operator. Yet another alternative is noted in BBM (p. 57).

With the definition (3.3) of operators at our disposal, we can readily appreciate the notion of continuity of modelling, inasmuch as it refers to the approximation of dispersive properties. For changes in the adopted approximation to $c(k)$, which in physical examples is generally a well-defined function, generate a continuous range of approximate dispersion operators. When the approximation to $c(k)$ is a (finite) polynomial, a differential operator is formed, but this case is certainly not exclusive. Indeed, real examples occur in which $c(k)$ is not twice differentiable at $k = 0$, so that a polynomial approximation is impossible, but the definition (3.3) is still perfectly serviceable. That is, the introduction of an approximation $c(k) = 1 - \alpha(k)$, with the function $\alpha(k)$ suitably tailored for small k but increasing without bound for $|k| \to \infty$, defines

$$\boldsymbol{L} \doteqdot \boldsymbol{I} - \boldsymbol{H},$$

where $\boldsymbol{H}$ is now a pseudo-differential operator with 'symbol' $\alpha(k)$. This interesting type of generalization, for which several physical applications are already known, will be considered further in § 3.4 and in § 5.

The basic mathematical difference between the KdV and BBM models can also be most readily appreciated by comparing the forms of the approximate dispersion relation for the respective linearized equations. For KdV it is (with $\gamma = 1$, as is always possible by rescaling x)

$$c = 1 - k^2,$$

corresponding to which the group velocity is

$$d(kc)/dk = 1 - 3k^2.$$

And for BBM it is

$$c = (1 + k)^{-1},$$

which, together with the group velocity, tends to zero as $|k| \to \infty$. Although these approximations coincide for small k (i.e. for long waves), they generate drastically different responses to short waves. The KdV model responds fiercely to any short-scale features with which it may happen to be presented (e.g. round-off errors in a numerical solution), and in this respect it is basically inexpedient as a long-wave model—notwithstanding other advantages that may fairly be claimed for it.

In contrast, the BBM model responds feebly to short waves. Thus it placidly does its job in simulating long-wave phenomena, without incidental behaviour of a troublesome kind. For the same reasons the qualitative theory of the equation is comparatively straightforward, and computing is also easy; so the model meets the criteria of practical expediency suggested at the beginning of this section. Weighing against this, a comparative insipidity of the mathematical aspects has to be recognized. The equation does not seem to offer the rich variety of resources, including exact solutions, that has been found for the KdV equation.

3.2. Theory of the KdV equation. An astonishing amount of effort has been expended on the study of the KdV equation in recent years, and some delightful results are available (e.g. see Lax [1968]). As already suggested, however, the KdV equation is in basic respects an inexpedient model for the various physical systems to which it is supposed to apply, and the many of its intriguing properties[6] revealed in recent work are irrelevant to these applications. Some reason for circumspection about physical relevancy arises particularly from the fact that most of the available theory depends crucially on the precise form of the equation, and so leaves open the important question of continuity of modelling with respect to varied specifications of the approximate dispersion operator. Nevertheless, the great ingenuity of recent accomplishments in KdV theory cannot be denied.

Because of its sensitivity to short-scale features, the equation is not easily amenable to numerical solution. Several successful numerical studies have been reported (e.g. by Vliegenthart [1971]), but they evidently testify to exceptional resourcefulness in devising stable finite-difference schemes. In effect, a drastic modification of the dispersive properties—as represented in the equation by the term u_{xxx}—has always to be incorporated in order to bring incidental short-scale features (round-off errors) under control.

Existence and regularity theory for the KdV equation is also difficult due to the short-wave properties of the equation—that is, to properties irrelevant to its role as a physical model. But various successful analyses have been accomplished, and the fundamental theory of the equation has been put into a fairly satisfactory state. For example, the existence of periodic solutions corresponding to periodic initial wave-forms was proved by Temam (1969), who used the method of parabolic regularization

[6] E.g. the existence of infinitely many invariants for C^∞ solutions that converge to zero, together with all their derivatives, as $x \to \pm\infty$.

as expounded in the book by Lions [1969] (this includes an account of Temam's work). The proof is not by any standard easy.

The problem of nonperiodic initial data on $(-\infty, \infty)$ is even more difficult, and Temam's analysis does not seem to be directly adaptable to it. An account of a new method of approach to the problem devised by my colleagues R. Smith and J. L. Bona is included in this volume. Their method is based on the strong existence and regularity results available for the BBM equation. A change of variables involving a parameter ε is shown to transform the BBM into a regularized version of the KdV equation, in which the regularizing term is $-\varepsilon u_{xxt}$, and the KdV results are hence obtained by investigating the limit $\varepsilon \to 0$.

These results can best be summarized by use of the notation $H^s(\boldsymbol{R})$ $[\equiv W_2^s(\boldsymbol{R})]$ for the space of square-integrable functions with s square-integrable generalized derivatives. Also, with respect to a finite time interval $[0, T]$, Banach spaces $\mathscr{H}_T^s$ are defined having the same relation to H^s as that of $\mathscr{L}_T$ to L_2 explained near the beginning of this section (i.e., $\mathscr{H}_T^1 = \mathscr{W}_T$ in the previous notation). For an initial wave-form $g(x) \in H^s(\boldsymbol{R})$ and forcing function $f(x, t) \in \mathscr{H}_T^s$, both with the same $s \geqq 3$, the KdV equation is shown to have a unique solution $u(x, t) \in \mathscr{H}_T^s$. Continuous dependence of the solution on g and f is also proved. Since $H^s(\boldsymbol{R}) \subset C^{s-1}(\boldsymbol{R})$ a classical, three times continuously differentiable solution is assured if $s \geqq 4$.

3.3. Theory of the BBM equation. Numerical solutions of the equation have been found by Peregrine [1966], and his work exemplifies that comparatively straightforward finite-difference schemes can be used successfully for this equation. A more or less exhaustive theory of existence, regularity and the continuous dependence of solutions on prescribed data was given in the paper already cited (BBM [1972], see also Bona and Bryant [1973]), and the result used at the beginning of this section to illustrate the Hadamard concept amounts to the main achievement of the paper. The method used to prove existence will now be outlined, it being enough for this purpose to consider the simpler case of the equation governing free wave propagation (i.e. $f \equiv 0$ in (3.1)).

The crux of the method is the following property of solutions. After multiplication by u, the equation can be arranged as

$$uu_t + u_x u_{xt} + (\tfrac{1}{2}u^2 + \tfrac{1}{3}u^3 - uu_{xt})_x = 0.$$

We provisionally assume that the three terms within the parentheses all vanish as $x \to \pm\infty$, and that the terms uu_t and $u_x u_{xt}$ are integrable. Hence, upon integrating the equation with respect to x between $-\infty$ and ∞, we may conclude that

$$E(u) = \int_{-\infty}^{\infty} (u^2 + u_x^2)\,dx = \text{const.} \tag{3.4}$$

This square root of this positive invariant is just the norm for the Sobolev space $W_2^1(\boldsymbol{R})$ considered earlier, and the assumption $E(u) < \infty$ implies that u is a continuous and uniformly bounded function of x on $-\infty < x < \infty$, converging to zero as $x \to \pm\infty$ [see (4.7) in § 4.2, also the penultimate paragraph of § 3.4]. Thus (3.4) indicates that, for initial data such that $E(g) < \infty$, a solution remains uniformly bounded for all $t > 0$.

[Note, incidentally, that another invariant is $F(u) = \int_{-\infty}^{\infty} (u^2 + \frac{1}{3}u^3)\,dx$ (see Benjamin [1972, p. 180]), and that the BBM equation may be expressed as $[\text{grad } E(u)]_t + [\text{grad } F(u)]_x = 0$. This form is common to all model equations of the present genre (cf. § 2.2). Its significance will be examined further in § 5.]

To put the existence problem into an amenable form, the partial differential equation first has to be transformed into an equivalent integral equation. Proceeding formally, one obtains

$$u_t = (\boldsymbol{I} - \partial_x^2)^{-1}(u + \tfrac{1}{2}u^2)_x,$$

and hence

$$u(x, t) = g(x) + \int_0^t \int_{-\infty}^{\infty} K(x - \xi)\{u(\xi, \tau) + \tfrac{1}{2}u^2(\xi, \tau)\}\,d\xi\,d\tau = \boldsymbol{A}u, \tag{3.5}$$

where the kernel function is

$$K(x) = \tfrac{1}{2}(\text{sgn } x)\,e^{-|x|}.$$

The proof is now developed in the following steps.

(1) *Local existence* ("*existence in the small*"). The classical Picard method of convergent iterations is used to show that, over a sufficiently small but finite time-interval, (3.5) has a continuous solution. More precisely, given that $g(x)$ is a continuous function with the bound $\sup_{x \in \boldsymbol{R}} |g(x)| \leqq b < \infty$, it is established that there exists a $t_0(b) > 0$ such that in the Banach space $\mathscr{C}_{t_0}$ (as defined earlier) the nonlinear integral operator $\boldsymbol{A}$ is a *contractive mapping* of a ball $\|v\|_{\mathscr{C}_{t_0}} \leqq R$ which includes the point g. Thus, by the contractive mapping principle, $\boldsymbol{A}$ has a unique fixed point u in the ball, and this is a continuous solution of the integral equation. The Picard sequence $\{v_n\}$ generated by the formula $v_n = \boldsymbol{A}v_{n-1}$, $v_1 = g$ converges strongly (in the norm of $\mathscr{C}_{t_0}$) to the limit u.

(2) *Regularity*. It is assumed that $g(x) \in C^2(\boldsymbol{R})$. Hence a "bootstrap" argument is used to show that the solution u has more regularity than membership of $\mathscr{C}_{t_0}$ alone implies (i.e. the fact that u is identical with Au is repeatedly exploited). The conclusion is that $u \in \mathscr{C}_{t_0}^{2,\infty}$ (i.e. the first two

x-derivatives and *all* t-derivatives of u are bounded and continuous functions on $\boldsymbol{R} \times [0, t_0]$). Thus u is shown to be a classical solution of the original initial-value problem.

(3) *Invariance of* $E(u)$. By use of the results (1) and (2) it is verified that, if $E(g) = E_0 < \infty$, then $E(u) = E_0$ throughout the interval $[0, t_0]$, as anticipated in (3.4).

(4) *Global extension.* Since $E(u) < \infty$ implies that $|u|$ has a uniform bound on $\boldsymbol{R}$, and since $E(u)$ is invariant over a time-interval depending only on such a bound, the preceding arguments can be applied repeatedly any number of times, thus extending without limit the time-interval over which a regular solution is assured.

Thus it is proven that, corresponding to any $g(x) \in C^2(\boldsymbol{R})$ with $E_0 < \infty$, there exists a function $u(x, t) \in \mathscr{C}_\infty^{2,\infty}$ satisfying the partial differential equation and the initial condition $u(x, 0) = g(x)$. Moreover, $E(u) = E_0$ at each $t \in [0, \infty)$.

An interesting aspect that comes to light in the step (2) is the following. Suppose that $g(x)$ is only *piecewise* twice differentiable (i.e. g'' has discontinuities). Then the same regularity argument shows that $u - g$ remains a C^2 function of x. Thus, if discontinuities in u_{xx} are introduced initially, they *do not propagate*, and remain undiminished in strength at their initial positions while the C^2 part of the solution evolves away from them. This weak form of response to discontinuities is in marked contrast to the corresponding behavior of solutions of the KdV equation. In the latter case the effects of an initial discontinuity in u_{xx} are instantly manifested over a range of x extending to $-\infty$.

3.4. Generalizations. With respect to applications, probably the most interesting generalizations of the KdV and BBM equations are those for systems with anomalous dispersion relations $c(k)$ that are not C^2 in a neighbourhood of $k = 0$. The modelling principle explained in § 3.2 then gives rise to a constant-coefficient pseudo-differential operator $\boldsymbol{H}$ in place of $-\partial_x^2$ in the KdV and BBM equations. An example arising in the theory of axisymmetric waves in a rotating fluid was noted in BBM (p. 74), as well as another that will be explained in § 5.2 below. For the general reasons given at the end of § 3.1, stronger existence and regularity results are possible for the BBM form of the generalized equation, and presumably the computing of solutions would also be much easier.

In Appendix 1 of BBM an existence theory is outlined applying to the initial-value problem for

$$(u + \boldsymbol{H}u)_t + u_x + uu_x = 0,$$

in the case that the symbol $\alpha(k)$ of $\boldsymbol{H}$ satisfies: (i) $\alpha(0) = 0$; (ii) $\alpha(k) = \alpha(-k)$;

(iii) $\alpha(k)$ is continuous and positive for $|k| > 0$; and (iv) $(1 + \alpha(k))^{-1} = O(1/k^2)$ for $|k| \to \infty$.

Conditions (i) and (ii) are inherent to any physically reasonable model. Condition (iv) implies that $\boldsymbol{H}$ is at least as strong as the corresponding operator $-\partial_x^2$ in the BBM equation, and in consequence of this the theory can be developed on essentially the same lines as for the third-order differential equation.

By extension of these arguments it is possible without much difficulty to cover the initial-value problem for

$$(u + \boldsymbol{H}u)_t + \{F(u)\}_x = f(x, t),$$

where (a) $F(u)$ is a C^1 function on $\boldsymbol{R}$; (b) $f(x, t) \in \mathscr{C}_T \cap \mathscr{L}_T$; and (c) $\alpha(k)$ satisfies conditions (i)–(iii) as above, but (iv) is relaxed to $(1 + \alpha(k))^{-1} = O(1/|k|^\beta)$ with $\beta > 1$. The initial wave-form $u(x, 0) = g(x)$ is suitably specified as an element of the Hilbert space, say $H^{\beta/2}$, whose norm is

$$\|g\|_{\beta/2} = \left\{\frac{1}{2\pi}\int_{-\infty}^{\infty} (1 + |k|^\beta)|\hat{u}|^2 \, dk\right\}^{1/2}$$

(note that $\beta = 2$ defines $H^1(\boldsymbol{R})$ as considered previously), and the solution $u(x, t)$ can be shown to remain in this space for each $t \in [0, T]$. Moreover, in the case $f \equiv 0$, it appears that, for all $t \geqq 0$,

$$\int_{-\infty}^{\infty} (u^2 + u\boldsymbol{H}u) \, dx = \text{const.} \geqq a\|u\|_{\beta/2}^2,$$

where a is a positive constant. A fact crucially useful for the existence and regularity theory of this generalized equation is that, if $\beta > 1$, the elements of $H^{\beta/2}$ are also continuous functions of x vanishing at infinity (this fact is immediately deducible from the Riemann-Lebesgue theorem for integrals: cf. context of (5.14) in § 5.3 below).

Finally, we note the possibility of a further generalization in which the pseudo-differential operator is allowed to depend on x. The only physically relevant models seem to be those where the operator is symmetric (i.e. in gradient form—see § 5.1(2)). Thus they must have the form $\boldsymbol{H}u = q\boldsymbol{K}u + \boldsymbol{K}(qu)$, where q is a fixed function of x and $\boldsymbol{K}$ is a constant-coefficient symmetric operator as considered previously.

4. Stability of steady waves

Here we return to the ideas discussed tentatively in § 2.3. Nonlinear dispersive systems of the general type now in question can usually sustain steadily translating waves, either solitary or periodic waves, and the appreciation of these special states of motion is essential to the explanation of many wave phenomena observed in practice. So questions about the stability of steady waves are no less important than questions of their existence as exact solutions of the dynamical equations. It was suggested in § 2.3, and is now to be considered more carefully in a particular example, that solitary waves are stable according to any realistic model, but periodic waves are generally unstable. The first of these propositions is easy to accept, being supported by a great deal of experimental evidence and by numerical solutions of the dynamical problem, but the second is considerably more contentious. Indeed, the value of much recent work concerning slowly modulated wave-trains seems to depend on assumptions of stability, and some of the mathematical issues remain obscure. Although the following discussion refers specifically only to the steady-wave solutions of the KdV equation, it deserves emphasis that present ideas should be applicable to other model equations (e.g. the generalizations treated in § 5).

We first need to recall the precise definition of stability, due originally to Liapunov and extended by many writers such as Movchan. This relies on a general notion of distance between possible states of a dynamical system at each instant—as represented by a vector or function. To be specific with regard to the present application, taking $\phi(x - Ct)$ to denote a steady wave which has to be compared with other possible motions $u(x, t)$, we have to adopt a measure of distance between ϕ and u qua functions of x. This *metric* is a scalar (functional) satisfying the axioms

(i) $d(\phi, u) \geqq 0 \quad (= \text{only if } \phi = u)$;

(ii) $d(\phi, u) = d(u, \phi)$;

(iii) $d(\phi, u) \leqq d(\phi, v) + d(v, u)$ (triangle inequality).

We shall in fact need different metrics for the initial conditions and the resulting motions, say d_1 and d_2 respectively. According to the Liapunov definition, ϕ is stable with respect to the particular metrics if the following condition is satisfied. For any positive number ε, however small, there exists another positive number δ_ε such that $[d_1(\phi, u)]_{t=0} \leqq \delta_\varepsilon$ implies $d_2(\phi, u) \leqq \varepsilon$ for all $t \geqq 0$. If this condition is found not to be satisfied, ϕ can be said to be unstable in the Liapunov sense.

This definition accords in most respects with intuitive concepts of stability and instability, and it is the only one that is really manageable in contexts like the present. But, of course, instability in this particular sense does not necessarily imply that small initial perturbations can result eventually in very large departures from the motion ϕ—that is, the wholly disruptive kind of situation that may be judged practically to constitute instability. The implication is just that motions neighbouring on ϕ are not amenable to precise control.

4.1. KdV solitary waves. In a recent paper (Benjamin [1972]) it has been demonstrated that the solitary-wave solutions of the KdV equation are stable, and the method of proof will now be outlined. It is noteworthy that the gist of the argument is exactly parallel to the tentative reasoning presented in § 2.3 with regard to the complete water-wave problem.

To find the class of solitary-wave solutions of

$$u_t + u_x + uu_x + u_{xxx} = 0, \tag{4.1}$$

we substitute

$$u = \phi(\bar{x}) \quad \text{with } \bar{x} = x - (1 + \bar{C})t,$$

and assume that the function ϕ and its second derivative ϕ'' vanish at infinity. Then one integration gives

$$\bar{C}\phi = \phi'' + \tfrac{1}{2}\phi^2. \tag{4.2}$$

For every (real) $\bar{C} > 0$, this equation has a solution vanishing at infinity, namely

$$\phi = 3\bar{C}\,\mathrm{sech}^2\{\tfrac{1}{2}\bar{C}^{1/2}(\bar{x} + a)\}, \tag{4.3}$$

where a is an arbitrary real constant. We shall henceforth use the symbol ϕ to mean some particular member of the class of solutions represented by (4.3), and for convenience of illustration we specify that $\bar{C} < 1$. (Note that the condition $\bar{C} \ll 1$ is warranted in respect of applications, since the validity of (4.1) as an approximate model for long waves in real systems depends on the magnitude of $|u|$ being small.)

On the assumption that solutions have a sufficient number of continuous and bounded derivatives vanishing as $x \to \pm\infty$, it may readily be confirmed that two invariants for (4.1) are

$$V(u) = \int_{-\infty}^{\infty} u^2\,dx, \qquad M(u) = \int_{-\infty}^{\infty} (u_x^2 - \tfrac{1}{3}u^3)\,dx. \tag{4.4}$$

And (4.2) is seen to be the Euler-Lagrange equation for the conditional variational problem $\delta M = 0$ for V fixed. Thus we have another, simpler example of the situation described at the end of § 2.2. Following the line of argument suggested in § 2.3, a demonstration of stability may be made by showing $M(\phi)$ to be a minimum of $M(u)$ under the constraint $V(u) = V(\phi)$, and hence considering the invariant $M(u) - M(\phi)$ as a "Liapunov functional."

4.2. Estimates for $M(u) - M(\phi)$. For the moment we focus attention on the variational problem, and the time-dependence of solutions is immaterial. Accordingly, u is treated as an arbitrary function of x alone, and in place of (4.3) we may consider simply

$$\phi = 3\bar{C}\,\mathrm{sech}^2(\tfrac{1}{2}\bar{C}^{1/2}x). \tag{4.5}$$

Writing $u = \phi + h$ and specifying

$$\Delta V = V(\phi + h) - V(\phi) = \int_{-\infty}^{\infty} (2\phi h + h^2)\,dx = 0,$$

we proceed to investigate

$$\Delta M = M(\phi + h) - M(\phi) = \int_{-\infty}^{\infty} \{(2\phi' h' - \phi^2 h) + h'^2 - \phi h^2 - \tfrac{1}{3}h^3\}\,dx.$$

Since ϕ satisfies (4.2), the linear terms in $\Delta M + \bar{C}\Delta V$ vanish, and so the constraint $\Delta V = 0$ implies

$$\begin{aligned} \Delta M &= \int_{-\infty}^{\infty} \{h'^2 + (\bar{C} - \phi)h^2\}\,dx - \int_{-\infty}^{\infty} \tfrac{1}{3}h^3\,dx \\ &= \delta^2 M + \delta^3 M, \quad \text{say}. \end{aligned} \tag{4.6}$$

We consider h as an element of the Sobolev space $W_2^1(\boldsymbol{R})$ with norm

$$\|h\| = \left\{\int_{-\infty}^{\infty} (h^2 + h'^2)\,dx\right\}^{1/2},$$

and make use of the inequality

$$\sup_{x\in\boldsymbol{R}} |h(x)| \leqq \|h\|/\sqrt{2} \tag{4.7}$$

(see Benjamin [1972, pp. 159 and 161]). Since ϕ is a nonnegative function and $\bar{C} < 1$, it follows from (4.6) that

$$\Delta M \leqq \|h\|^2 + \|h\|^3/(3\sqrt{2}), \tag{4.8}$$

which shows that positive values of ΔM can be kept arbitrarily small by specifying $\|h\|$ to be small enough. We shall take $d_1(u, \phi) = \|h\|$ in applying the Liapunov criterion of stability.

The task of deriving a useful lower bound for ΔM is much more difficult. We first note that since the functionals are translation invariant, perturbations $\phi + h$ that are just translations of ϕ obviously give $\Delta V = \Delta M = 0$. Moreover, a perturbation that is another solitary wave initially close to ϕ will eventually become far separated from ϕ, however small the difference in the speed $\bar{C}$ of translation. It would therefore be useless to adopt $d_2(u, \phi) = \|u - \phi\|$ as the second metric for the criterion of stability. We need in fact to recognize that stability of a solitary wave can only be established with respect to its shape (not also with respect to its position at all times), and to discriminate the stability problem in this respect the following device may be used.

Let τ stand for the group of translations along the x-axis, that is, $\tau u(x) = u(x - \xi)$ where ξ ranges over all real numbers. From $W_2^1(\boldsymbol{R})$ a quotient space S may be formed by identifying the translations of each function $u \in W_2^1(\boldsymbol{R})$. That is, each element of S is an equivalence class of functions that are translations of each other; and in taking a particular function $u(x)$ to represent an element of S, we can arbitrarily relocate the origin of x. For example, the solitary-wave solution (4.3) is for all t the same in S and is representable by (4.5). A metric for S is

$$d(u, v) = \inf_{\tau} \|\tau u - v\|$$

(cf. Benjamin [1972, p. 160]), and we shall adopt this as d_2.

The analysis in the cited paper proceeds by estimating the contributions $\delta^2 M$ made separately by the even and odd components of h. The condition $\Delta V = 0$ serves primarily to restrict the even component of h, and the extra constraint imposed by consideration of u as an element of S applies only to the odd component. In fact, it is specified that every perturbation $u(x)$ be relocated as $u(x - a)$, where a is determined by

$$\text{(4.9)} \qquad \int_{-\infty}^{\infty} \{u(x - a) - \phi(x)\}^2 \, dx = \inf_{\xi \in \boldsymbol{R}} \int_{-\infty}^{\infty} \{u(x - \xi) - \phi(x)\}^2 \, dx.$$

Then, with $h(u) = u(x - a) - \phi(x)$, it follows that $\int_{-\infty}^{\infty} h(x)\phi'(x) \, dx = 0$, which, since $\phi'(x)$ is an odd function, is a condition on only the odd component $g(x) = \frac{1}{2}\{h(x) - h(-x)\}$. The usefulness of the estimates obtained on the basis of this particular mode of comparison between ϕ and u (considered as elements of S) owes to the fact that, with h thus delimited, $\|h\| \geqq d_2(u, \phi)$ and $\|h\|$ varies continuously with t when u is a solution of (4.1).

A lower bound for $\delta^2 M$, subject to the aforementioned conditions on h, is obtained by means of spectral theory (loc. cit., pp. 164–168), and an alternative derivation pased on methods of the calculus of variations is also given (loc. cit., Appendix A). The details are complicated and will be passed over here. The result is

$$\delta^2 M \geqq \frac{1}{4} \int_{-\infty}^{\infty} (h'^2 + \bar{C} h^2) \, dx - \frac{2}{5} \bar{C}^{1/4} \|h\|^3,$$

and from this, again using the inequality (4.7) and the assumption $\bar{C} < 1$, we may conclude from (4.6) that

(4.10) $$\Delta M \geqq \tfrac{1}{4}\bar{C}\|h\|^2 - \tfrac{1}{3}b\|h\|^3,$$

where $b = 6\bar{C}^{1/4}/5 + 1/\sqrt{2}$. It is evident from this estimate that $M(\phi)$ is a conditional minimum.

4.3. Conditional stability of solitary waves. From the results (4.8) and (4.10), we may infer that the solitary-wave solution ϕ is stable, subject to the condition that perturbations of the solution initially (and hence permanently) satisfy $V(u) = V(\phi)$. The argument goes as follows (cf. loc. cit., p. 169).

Let A be the positive root of

$$A^2 + A^3/3\sqrt{2} = \bar{C}^3/96b^2.$$

Then the initial condition

(4.11) $$[d_1(u, \phi)]_{t=0} = \|h\|_{t=0} \leqq A$$

implies, according to (4.8), that $\Delta M \leqq \bar{C}^3/96b^2$, and this inequality applies for all $t \geqq 0$ since ΔM is invariant. Hence (4.10) shows that, if $\|h\|$ varies continuously with t (as may be confirmed),[7] then $\|h\| \leqq \bar{C}/4b$, and correspondingly

$$\Delta M \geqq \tfrac{1}{6}\bar{C}\|h\|^2.$$

(Remember that here the functions $h\,(= u - \phi)$ have been adjusted according to (4.9), and that consistently with this adjustment $\|h\| \geqq d_2(u, \phi)$.) Thus we have finally that

$$\bar{C}^3/96b^2 \geqq (1 + A/3\sqrt{2})[d_1(u, \phi)]^2_{t=0} \geqq \Delta M$$

implies

$$\tfrac{1}{6}\bar{C}[d_2(u, \phi)]^2 \leqq \Delta M \quad \text{for } t \geqq 0.$$

Since ΔM is independent of t, this evidently establishes that ϕ is (conditionally) stable with respect to the considered metrics.

4.4. General stability of solitary waves. It is now a simple matter to show that the solitary-wave solution is generally stable, for perturbations free from the previous restriction $V(u) = V(\phi)$ (cf. loc. cit., p. 170). Allowing that $V(u)$ may differ from $V(\phi)$, we define ϕ_V as the solitary wave for which $V(\phi_V) = V(u)$, and which is initially centred on the same point as ϕ. Hence, supposing that $[d_1(u, \phi)]_{t=0} = \|u - \phi\|_{t=0} \leqq \delta$, we may readily confirm that

---

[7] This property was presumed in the cited paper, but it has been verified by my colleague Dr. J. L. Bona.

$$[d_1(\phi, \phi_V)]_{t=0} = \|\phi - \phi_V\|_{t=0} = d_2(\phi, \phi_V)$$

has a δ-determined upper bound (i.e. a bound that is reducible towards zero in step with δ).

From the triangle inequality $d_1(u, \phi_V) \leqq d_1(u, \phi) + d_1(\phi, \phi_V)$, it follows that the initial value of $d_1(u, \phi_V)$ has a δ-determined upper bound. Hence, using the conclusion established in § 4.3, we may obtain a δ-determined upper bound for $d_2(u, \phi_V)$ which applies for all $t > 0$ and all the possible ϕ_V. Finally, the triangle inequality for the metric d_2 is used:

$$d_2(u, \phi) \leqq d_2(u, \phi_V) + d_2(\phi_V, \phi).$$

Since both terms on the right-hand side have been shown to have δ-determined upper bounds holding for all $t > 0$, it is thus shown that the solution ϕ is stable in the general sense.

4.5. Cnoidal waves. The class of periodic steady solutions of (4.1) is also well known (Korteweg and de Vries [1895], Lamb [1932, § 253]). After the substitution $u = \phi(\bar{x})$, where now ϕ is supposed to be a periodic function of $\bar{x} = x - (1 + \bar{C})t$, equation (4.2) is again obtained by integrating (4.1) if the constant of integration is arbitrarily put equal to zero (as is justifiable; see remark below (5.1) in next section). In terms of the Jacobian elliptic function cn, the periodic solutions of (4.2) may be written

$$\phi = a + (b - a)\mathrm{cn}^2(\beta\bar{x}; k), \tag{4.12}$$

with $\beta^2 = (b + l)/12$, $k^2 = (b - a)/(b + l)$, $l = a + b - 3\bar{C} = ab/(a + b)$. It follows from these definitions that $0 < a < 2\bar{C} < b < 3\bar{C}$, and the period (wavelength) is

$$\lambda = \frac{4\sqrt{3} \cdot K(k^2)}{b + l} > \frac{2\pi}{\bar{C}^{1/2}}. \tag{4.13}$$

If, in place of (4.4), functionals $V_\lambda(u)$ and $M_\lambda(u)$ are defined as the

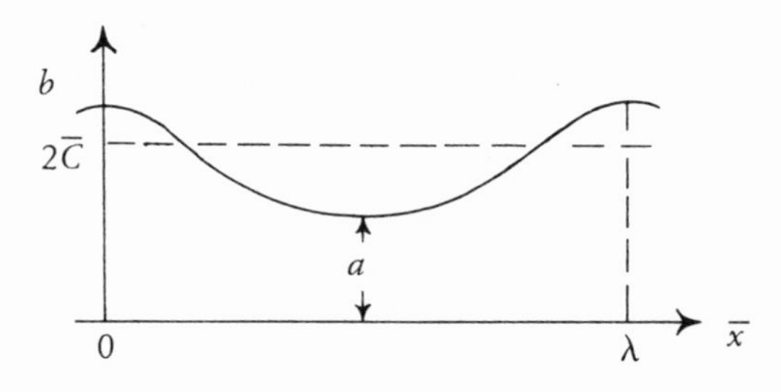

FIGURE 3

respective integrals over one period, (4.2) may now be interpreted as the Euler-Lagrange equation for the conditional variational problem $\delta M_\lambda = 0$ for V_λ fixed, where only functions with period λ compete for the stationary value of M_λ. Unlike the previous case, however, there is now a trivial nonzero solution of the problem, namely the constant

$$\phi_0 = 2\bar{C}_0 = (V_\lambda/\lambda)^{1/2}.$$

For this extremal the conditional second variation of M_λ is

$$\delta^2 M_\lambda = \int_0^\lambda (h'^2 - \bar{C}_0 h^2)\, dx$$

with $\int_0^\lambda h\, dx = 0$. Hence the Wirtinger inequality gives

$$\delta^2 M_\lambda \geqq \left\{ \left(\frac{2\pi}{\lambda} \right)^2 - \bar{C}_0 \right\} \int_0^\lambda h^2\, dx.$$

Thus, if $\bar{C}_0 = \frac{1}{2}(V_\lambda/\lambda)^{1/2} < (2\pi/\lambda)^2$, the second variation is positive, and it is in fact easy to confirm that $M(\phi_0)$ is then a conditional minimum. Moreover, ϕ_0 is then the only nonzero extremal, as might be expected from the inequality (4.13).

On the other hand, if $\bar{C}_0 > (2\pi/\lambda)^2$, the second variation of M_λ is negative for $h = \varepsilon \cos (2\pi x/\lambda)$, and so evidently $M(\phi_0)$ is not a conditional minimum in this case. It can in fact be shown (by methods of the classical calculus of variations, or by Hilbert-space methods such as used in § 5.4 below) that the solution (4.12), with appropriate a, b and $\bar{C}$ such as to make λ the fundamental period, then realizes the absolute minimum of M_λ for a given V_λ [i.e. given $V_\lambda > 64\pi^4/\lambda^3$ so as to make $\bar{C}_0 > (2\pi/\lambda)^2$]. Furthermore, if variations that are just translations of ϕ are excluded from competition, say by the device explained in § 4.2, the functional $M_\lambda(u) - M_\lambda(\phi)$ can be bounded positively from below in terms of $\|u - \phi\|$. Since this functional is invariant for solutions of the KdV equation (4.1) having period λ, the arguments applied above to solitary waves now show that the periodic solution (4.12) is *stable with respect to perturbations with the same period.*

The demonstration of stability fails, however, if we allow perturbations with a sufficiently longer period. Consider, for example, the case of perturbations with period 2λ. Evidently, $M_{2\lambda}(\phi) = 2M_\lambda(\phi)$ is a stationary value but is not the absolute minimum of $M_{2\lambda}(u)$ for a given $V_{2\lambda} = 2V_\lambda$, because this minimum will be realized by the solution of (4.2) whose fundamental period is 2λ. In fact, as the following simple argument shows, $M_{2\lambda}(\phi)$ is *not* a conditional minimum.

The conditional second variation of $M_{2\lambda}$ about the extremal ϕ is

$$\delta^2 M_{2\lambda} = \int_0^{2\lambda} \{h'^2 + (\bar{C} - \phi)h^2\}\, dx = -\int_0^{2\lambda} h\mathscr{L}h\, dx, \tag{4.14}$$

where $\mathscr{L}h = h'' + (\phi - \bar{C})h$, and h is required to satisfy

$$\int_0^{2\lambda} \phi h\, dx = 0. \tag{4.15}$$

Locating the origin of x so that $\phi(x)$ is an even function, we consider the Sturm-Liouville problem

$$\mathscr{L}\zeta = \mu\zeta, \qquad \zeta(x) = \zeta(x + 2\lambda),$$

in the case of *odd* eigenfunctions ζ [i.e. $\zeta(0) = \zeta(2\lambda) = 0$, $\zeta'(0) = \zeta'(2\lambda)$]. Differentiation of the equation (4.2) satisfied by ϕ shows that ϕ' is an eigenfunction, corresponding to $\mu = 0$, and ϕ' has three zeros in the open interval $(0, 2\lambda)$. Hence, according to elementary Sturm–Liouville theory, there is an eigenfunction ζ_1 with a single zero in $(0, 2\lambda)$ and for which the eigenvalue $\mu_1 > 0$. Being an odd function, ζ_1 satisfies the isoperimetric condition (4.15) and, when substituted in (4.14), ζ_1 gives a negative value of the second variation. This shows that $M_{2\lambda}(\phi)$ cannot be a conditional minimum.

It thus appears there is no hope of showing by present means that the cnoidal-wave solution ϕ is stable for perturbations with twice its period, and the possiblity is indicated that ϕ may be unstable in the Liapunov sense. The problem of proving whether or not this is true presents formidable difficulties, however, and a solution will not be attempted here. The following comments are included merely to intimate current work which will be reported later.

4.6. Instability. Whereas the property of $M(\phi)$ being a conditional minimum of the invariant functional M is crucial to the demonstration of stability for solitary waves (and, in respect of M_λ, for cnoidal waves when perturbed without changing the period of the motion), the absence of this property does not by itself, of course, imply that the respective steady wave is unstable. For example, the constant solution $\phi_0 = 2\bar{C}_0$ of (4.2) does not give a conditional minimum among perturbations with period $\lambda > 2\pi/\bar{C}_0^{1/2}$, but this solution appears to be stable for all periodic perturbations. (For, obviously, $h = u - \phi_0$ satisfies the form of the KdV equation with $(1 + 2\bar{C}_0)$ replacing 1 as the coefficient of the second term in (4.1), and there is no evidence of the zero solution of the KdV equation being unstable).

On the other hand, considering the physical meaning of the solution ϕ_0, we appreciate that it is unstable in a certain intuitive sense. We recall that (4.2) was obtained by substituting $u = \phi\{x - (1 + \bar{C})t\}$ in (4.1), integrating and putting the constant of integration equal to zero. Thus, the solution ϕ_0 is significant as the asymptote as $\bar{x} \to -\infty$ of a solution representing a long surge which advances with speed $1 + \bar{C}$ into a region at rest (see Figure 2 on p. 7). Now suppose that the wave front is sufficiently far along the x-axis so that, in a neighbourhood of the fixed station $x = 0$, we have $u = \phi_0$ and $u_x = u_{xx} = u_{xxx} = 0$ with negligible error. Let $V_+(u)$ and $M_+(u)$ be defined by integrals like (4.4), but with $x = 0$ as the lower limit of integration. From (4.1) it may easily be verified (cf. Benjamin [1972, p. 156]) that

$$\begin{aligned} dV_+/dt &= \phi_0^2 + \tfrac{2}{3}\phi_0^3, \\ dM_+/dt &= -\tfrac{1}{3}\phi_0^3 - \tfrac{1}{4}\phi_0^4. \end{aligned} \tag{4.16}$$

But, if the region where $u = \phi_0$ were to extend in the $+x$ direction at the wave speed $1 + \bar{C}_0 = 1 + \frac{1}{2}\phi_0$, while (as may be assumed) the solution and its derivative remained bounded at the wave-front, then we would have eventually

$$\begin{aligned} dV_+/dt &= (1 + \tfrac{1}{2}\phi_0)(\phi_0^2) = \phi_0^2 + \tfrac{1}{2}\phi_0^3, \\ dM_+/dt &= (1 + \tfrac{1}{2}\phi_0)(-\tfrac{1}{3}\phi_0^3) = -\tfrac{1}{3}\phi_0^3 - \tfrac{1}{6}\phi_0^4. \end{aligned} \tag{4.17}$$

The differences between the expressions (4.16) and (4.17), respectively $\phi_0^3/6$ and $-\phi_0^4/12$, indicate that local mean values of u^2 and $u_x^2 - u^3$ respectively increase and decrease in regions behind the wave-front, relative to the values for $u = \phi_0$. Hence it may be concluded that undulations develop approximating to cnoidal waves. This behavior is in fact well known from numerical solutions of the KdV equation (e.g. Vliegenthart [1971, p. 143]).

Another view of the matter is given by putting $u = \phi_0 + h(\bar{x}, t)$, $\bar{x} = x - (1 + \bar{C}_0)t$, in (4.1) and linearizing in h. Thus we get

$$h_t + \bar{C}_0 h_x + h_{xxx} = 0.$$

Solutions with the property $h \to 0$ for $\bar{x} \to -\infty$ may be supposed to represent, approximately, the growth of perturbations far behind the wave-front, and it is reasonable not to require them to be bounded for $\bar{x} \to \infty$ (since nonlinear effects will "take over" at the wave-front). For example, consider $h = a \exp(\kappa t - ik\bar{x})$ with $k = \bar{C}_0^{1/2} + i\varepsilon$, $0 < \varepsilon < (2\bar{C}_0)^{1/2}$. Then we have

$$\mathrm{Re}\,(\kappa) = 2\bar{C}_0\varepsilon - \varepsilon^3 > 0.$$

Thus, at a point $\bar{x} =$ const. far behind but travelling with the wave-front, the perturbation h grows exponentially with time.

We now turn to the question whether the periodic steady wave $\phi(\bar{x})$ given by (4.12) may be unstable in the Liapunov sense. Putting $u = \phi(\bar{x}) + h(\bar{x}, t)$ in (4.1), we obtain as the equation for h,

$$h_t + (\mathscr{L}h)_x + hh_x = 0, \tag{4.18}$$

where $\mathscr{L}$ is the second-order differential operator defined below (4.14), with coefficient $\{\phi(\bar{x}) - \bar{C}\}$. There is some evidence pointing to the possibility that the zero solution of (4.18) may be unstable, but proof is elusive. One natural approach is to consider whether the linearized equation

$$h_t + (\mathscr{L}h)_x = 0 \tag{4.19}$$

has a solution, periodic in $\bar{x}$ with period $N\lambda$ (N an integer > 1), that becomes unbounded as $t \to \infty$. Suppose that for the third-order, nonself-adjoint eigenvalue problem

$$(\mathscr{L}\xi)' = \sigma\xi,$$
$$\xi(\bar{x}) = \xi(\bar{x} + N\lambda)$$

it could be shown that there exists at least one eigenfunction, say $\xi_1(\bar{x})$, for which $\mathrm{Re}(\sigma_1) < 0$. Then the solution $h = a\xi_1(\bar{x})\, e^{-\sigma_1 t}$ of (4.19) would obviously preclude the Liapunov criterion of stability from being satisfied for the linearized version of the perturbed system. This conjecture remains to be settled one way or the other in the particular case of KdV cnoidal waves; but definite evidence of the instability of periodic steady waves has come to light in studying linearized perturbation equations, akin to (4.19), derived for alternative long-wave models with the same formal status as the KdV equation.

Note that the present question of instability is considerably more demanding than in analogous cases of dynamical systems with a *finite* number of degrees of freedom. Concerning the latter it is well known that instability of a linearized approximation [when the nonlinear effects are $o(\|h\|)$] implies the instability of the exact system. But the same inference cannot generally be drawn for systems governed by partial differential equations, even though such evidence at least gives some likelihood to the possibility of instability.

Finally, it needs to be remarked that all the preceding ideas can be adapted to the study on the solitary-wave and cnoidal-wave solutions of the BBM equation (see Benjamin [1972, Appendix B]). In certain respects, particularly because better-behaved operators are involved, the BBM equation seems to be a more expedient model for investigating the possibility of cnoidal waves being unstable.

5. Steady solutions of generalized evolution equations

In § 3.2 it was discussed how nonlinear systems with general dispersive properties may be modelled by equations involving abstract linear operators. For such equations we now consider the problem of establishing the existence of solutions representing waves of permanent form. This problem is different in character from the initial-value problem for the respective equations, but the methods of functional analysis are again needed to develop any definite theory. According to the type of model equation adopted, the mathematical problem may be put into various forms, so we first review the possibilities.

5.1. Possible models.

1. *Equations incorporating an exact dispersion operator.* One rational model pointed out in § 3.2 is the equation

$$u_t + (\boldsymbol{L}u)_x + uu_x = 0,$$

in which the first two terms on the left-hand side recover the exact result given by linearized theory (for unidirectional propagation), and the third derives from a long-wave approximation of nonlinear effects. In terms of Fourier transforms [denoted $\hat{u}(k) = \mathscr{F}u(x)$], the operator $\boldsymbol{L}$ is defined by $(\boldsymbol{L}u)\hat{} = c(k)\hat{u}$, where $c(k)$ is the exact phase-velocity function according to linearized theory. (A corresponding definition for periodic functions of x will be explained in § 5.2 below.)

Assuming a solution in the form $u = \phi(x - Ct)$, we obtain

$$C\phi_x = (\boldsymbol{L}\phi)_x + \phi\phi_x,$$

and hence

$$C\phi = \boldsymbol{L}\phi + \tfrac{1}{2}\phi^2. \tag{5.1}$$

No generality is lost by taking the constant of integration to be zero, because the constant can be absorbed into a new variable satisfying an equation in the same form as (5.1).[8]

Usually $c(k) \to 0$ as $|k| \to \infty$, and so $\boldsymbol{L}$ is a smoother operator than identity. In other words, considering $\boldsymbol{L}u$ as a convolution $(K * u)$, we have that the kernel $K(x)$ (the inverse Fourier transform of $c(k)$) is well behaved; typically $c(k) \in L_2$ and therefore $K(x) \in L_2$, which implies that $\boldsymbol{L}$ is a continuous transformation from L_2 into C. The place taken by this smooth operator in (5.1), separately from the nonlinear term, makes the equation especially difficult to handle. Although an existence theory for (5.1) seems

[8] Note that throughout this section C will stand for the *absolute* wave velocity denoted by $1 + \bar{C}$ in § 4.

possible by means of variational methods, it would be considerably more complicated than for the alternative models considered below.

Unlike the alternative models, (5.1) ceases to have a nontrivial solution when the parameter C is sufficiently large. Consider, for example, the possibility of a solitary-wave solution in the case that $\boldsymbol{L}$ is a positive operator (i.e. $K \geqq 0$ so that $\boldsymbol{L}u \geqq 0$ if $u \geqq 0$). It is easily seen that in this case a solitary-wave solution must be positive. A crude estimate of the upper limit for C may then be deduced by noting from (5.1) that, since ϕ must vanish with $\boldsymbol{L}\phi$, therefore

$$\phi = \bar{C} - (\bar{C}^2 - 2\boldsymbol{L}\phi)^{1/2},$$

which shows that $\phi \leqq C$. But the normalization $c(0) = 1$ implies that

$$(C - 1)\int \phi \, dx = \frac{1}{2}\int \phi^2 \, dx,$$

hence $C > 1$ and $\sup \phi > 2(C - 1)$. It follows that $C < 2$.

It was suggested in § 3.2 that a more satisfactory model is

$$u_t + \boldsymbol{L}(u_x + uu_x) = 0.$$

The spectrum (symbol) $c(k)$ of the operator $\boldsymbol{L}$ is generally a positive function, and this implies that $\boldsymbol{L}$ can be "split" in the form $\boldsymbol{L}^{1/2}(\boldsymbol{L}^{1/2})^*$. Hence the problem can be put into a form that will be explained in (3b) below, and the theory of steady waves can proceed on the same lines as will be discussed with reference to the later example.

2. *Pseudo-differential equations.* As was also noted in § 3.4, a generalization of the BBM equation is

$$u_t + (\boldsymbol{H}u)_t + u_x + uu_x = 0, \tag{5.2}$$

where $\boldsymbol{H}$ is a pseudo-differential operator whose symbol $\alpha(k)$ is an even positive function satisfying $\alpha(0) = 0$ and $\alpha(k) \to \infty$ for $|k| \to \infty$. In this case the equation of steady waves is

$$C(\phi + \boldsymbol{H}\phi) = \phi + \tfrac{1}{2}\phi^2. \tag{5.3}$$

The corresponding generalization of the KdV equation has $-\boldsymbol{H}u_x$ in place of the second term of (5.2), and so in the counterpart of (5.3) the left-hand side is $C\phi + \boldsymbol{H}\phi$. In addition to the greatly increased difficulty of the initial-value problem, the theory of steady waves for the KdV generalization also presents difficulties avoided by (5.3). The applications of variational methods that will be explained presently depend crucially on the manner in which the parameter C enters the problem, and for this reason they do not appear to be directly adaptable to this alternative type of model.

We note that two nonlinear invariants for (5.2) are

$$V = \int (u^2 + u\boldsymbol{H}u)\,dx, \qquad (5.4)$$
$$W = \int (u^2 + \tfrac{1}{3}u^3)\,dx.$$

[The invariance of V is deducible immediately upon multiplying (5.2) by u, integrating and using the fact that $\boldsymbol{H}$ is a symmetric operator. To show that W is invariant, put $u_t = w_x$. Then it is seen that

$$\frac{dW}{dt} = \int w_x(2u + u^2)\,dx = -2\int w(u_x + uu_x)\,dx$$
$$= \int 2w(w_x + \boldsymbol{H}w_x)\,dx = 0,$$

since

$$\int w\boldsymbol{H}w_x\,dx = \int w_x\boldsymbol{H}w\,dx \qquad \text{(symmetry of } \boldsymbol{H})$$
$$= -\int w_x\boldsymbol{H}w\,dx \quad \text{(integration by parts).]}$$

The variational significance of such invariants with regard to steady waves has already been discussed in §§ 2 and 4, and we shall consider presently how the same ideas may be brought to bear on questions of existence.

First we need a precise definition of the gradient of a functional. Suppose the functional $F(u)$ is defined on some open subset Ω of a Hilbert space $\mathscr{H}$, whose inner product is written (u, v) and the norm $\|u\| = (u, u)^{1/2}$. $F(u)$ is said to have a linear Gateaux differential if, for all $u \in \Omega$ and all $h \in \mathscr{H}$, the formula

$$\lim_{s\to 0}\left\{\frac{F(u + sh) - F(u)}{s}\right\} = (\boldsymbol{G}u, h)$$

defines the right-hand side as a linear functional of h. Then the operator $\boldsymbol{G} = \operatorname{grad} F$ may properly be called the gradient of F. The gradient is said to be strong if $F(u + sh) - F(u) - s(\boldsymbol{G}u, h)$ is $o(s\|h\|)$.

Now, although V and W are not defined over any *open* subset of L_2, let us for the moment loosely interpret the foregoing definition as referring to elements of L_2 for which V and W do exist. Then we see that (4.3) is formally equivalent to

$$C \operatorname{grad} V(\phi) = \operatorname{grad} W(\phi). \qquad (5.5)$$

A similar (equally loose!) interpretation of the gradient has already been used in § 2.2, leading to the equation (2.10) comparable with (5.5).

It has already been pointed out that this result may be interpreted as the Euler-Lagrange necessary condition for an extremal of the isoperimetric problem: $W = \max$ or min for V fixed. So, if a nontrivial function $\phi(x)$ were known to realize a maximum of W for $V = V(\phi)$, it could be concluded that this function is a solution of (5.3). This simple observation illustrates the gist of a possible existence theory using variational methods, but, of course, for the reasons indicated in introducing (5.5), the present interpretation is ill-suited to the task of establishing a conditional maximum of W.

To make progress, one may consider the real Hilbert H with inner product

$$(u, v)_H = \int (uv + u\boldsymbol{H}v)\,dx = \int (uv + v\boldsymbol{H}u)\,dx = \frac{1}{2\pi}\int \{1 + \alpha(k)\}\hat{u}\bar{\hat{v}}*\,dk.$$

The last equality follows from Parseval's theorem (note $\bar{\hat{v}}*$ denotes the complex conjugate of $\hat{v}$); the corresponding expression for periodic functions will be given in § 5.3 below. Thus $V(u) = \|u\|_H^2$, and the formula (5.4) shows that in this space the (strong) gradient of $\frac{1}{2}V$ is just $\boldsymbol{I}$ (the identity operator). With the new interpretation, equation (5.5) still applies, but it stands instead for

$$C\phi = (\boldsymbol{I} + \boldsymbol{H})^{-1}(\phi + \tfrac{1}{2}\phi^2), \tag{5.6}$$

on the assumption that the right-hand side of this equation is defined.

By adaptation of known arguments (e.g. see Berger and Berger [1968, § 4.3]), an existence theory may be completed on the lines just indicated, at least for the case of periodic solutions. For this case in fact, if the positive function $\alpha(k)$ is such that $\{1 + \alpha(k)\}^{-1} = O(|k|^{-\beta})$ with $\beta > 1$ for $|k| \to \infty$, it may be shown that the functional $W(u)$ is weakly continuous (vide infra) in H. It follows that W achieves a maximum in the closed ball $V(u) \leqq R < \infty$, and that the extremal point must be on the spherical boundary $V(u) = R$. Hence the existence of a *weak* solution of (5.5) may be inferred (cf. Berger and Berger), and the remaining task—which seems the hardest part of the analysis in principle—is to establish sufficient regularity of this solution. Although this line of argument appears to be more or less standard in the mathematical literature nowadays, and is probably the most stylish way of dealing with the present problem, some of the logical details are very delicate. In the writer's view, the essential issues of the problem can be understood rather more easily in an alternative approach that is to be introduced in 3(b) below and carried through in § 5.4. This proceeds on slightly more old-fashioned lines, using ideas concerned particularly with the space L_2 such as are expounded in the book on variational methods by Vainberg [1964].

3. *Compact-operator equations.* (a) Fixing the number $\lambda = C - 1 > 0$, we may recast (5.3) in the form

$$\phi = \{\lambda \boldsymbol{I} + \boldsymbol{H}\}^{-1}(\tfrac{1}{2}\phi^2) = \tfrac{1}{2}\boldsymbol{A}\phi^2, \tag{5.7}$$

where the linear operator $\boldsymbol{A}$ is defined by $(\boldsymbol{A}u)^\wedge = \hat{u}/\{\lambda + \alpha(k)\}$. In the case that $\boldsymbol{A}$ turns out to be a positive operator [i.e., in an integral representation $\boldsymbol{A}u = (K * u)$, the kernel function $K(x) \geqq 0$: note that K is the inverse Fourier transform of $\{\lambda + \alpha(k)\}^{-1}$], the operation on the right-hand side of (5.7) will take a cone of nonnegative functions into itself. Accordingly, use may be made of certain fixed-point theorems concerning mappings of cones in topological spaces (Krasnosel'skiĭ [1964], Benjamin [1971, §§ 2.2 and 3.3]). For the relevant theorems to be applicable, the mapping is also required to be completely continuous (compact), and in the present application to periodic waves a suitable Banach space (e.g. C) can generally be specified for which this requirement is satisfied. It has been found fairly straightforward to complete an existence theory for periodic waves on these lines (in fact based on Theorem B, p. 609, given in Benjamin [1971]), and a feature of the method perhaps worth mentioning is that Leray-Schauder degree theory serves to exclude the two trivial solutions $\phi \equiv 0$ and $\phi = 2\lambda$ of (5.7); but the details do not seem worth presenting here.

An interesting generalization of the relevant cone theorem has been established by my colleagues J. L. Bona and D. K. Bose, who have applied it successfully to the more difficult problem of proving that (5.7) has a solitary-wave solution. An account of their work is included in this volume. The particular difficulty of the problem owes to the fact that, as a positive mapping of any useful cone in a Banach space of functions defined on the *whole real axis*, the operator $\boldsymbol{A}$ is not completely continuous (cf. remarks in § 2.1(2)). The difficulty is obviated by considering the solution as an element of a suitable metric space (Fréchet space) "wider" than any relevant normed space, and by choosing a cone in this space such that, as well as being positive, the mapping $\boldsymbol{A}u^2$ of the cone is completely continuous in an appropriate sense respective to the coarser topology.

(b) *Simpler version of variational problem.* Since $1 + \alpha(k)$ is a positive function, the inverse operator $(\boldsymbol{I} + \boldsymbol{H})^{-1}$ appearing in (5.6) may be split in the form

$$(\boldsymbol{I} + \boldsymbol{H})^{-1}u = \boldsymbol{B}(\boldsymbol{B}^*u),$$

where $\boldsymbol{B}$ and $\boldsymbol{B}^*$ both have the symbol $\gamma(k) = \{1 + \alpha(k)\}^{-1/2}$, but are to be interpreted as adjoint operators in a sense explained presently. Evidently $\boldsymbol{B}u \equiv 0$ implies $u \equiv 0$ and vice versa. Hence, substituting $u = \boldsymbol{B}v$ and applying the operator $\boldsymbol{B}^*$ to equation (5.2), we see that the equation is equivalent to

$$v_t + [\boldsymbol{B}^*\{\boldsymbol{B}v + \tfrac{1}{2}(\boldsymbol{B}v)^2\}]_x = 0. \tag{5.8}$$

This may also be expressed as

$$\partial_t[\text{grad } V(v)] + \partial_x[\text{grad } W(v)] = 0,$$

where now, in place of (4.4),

$$V(v) = \int v^2\, dx, \qquad W(v) = \int\left\{(\boldsymbol{B}v)^2 + \frac{1}{3}(\boldsymbol{B}v)^3\right\} dx, \tag{5.9}$$

and the gradient is taken in L_2. As thus defined, V and W are invariants for the evolution equation (5.8).

The corresponding equation for steady waves is

$$C\psi = \boldsymbol{B}^*\{\boldsymbol{B}\psi + \tfrac{1}{2}(\boldsymbol{B}\psi)^2\}, \tag{5.10}$$

i.e.

$$C \text{ grad } V(\psi) = \text{grad } W(\psi), \tag{5.10'}$$

where $\boldsymbol{B}\psi$ is identifiable with the function ϕ considered in 2 above. The advantage of this formulation is that, with easily prescribable conditions on the operators $\boldsymbol{B}$ and $\boldsymbol{B}^*$, we can pose the existence problem in L_2. For instance, let $\boldsymbol{B}$ and $\boldsymbol{B}^*$ be defined as mappings $L_2 \to L_2$, and also in the sense

$$\boldsymbol{B}: L_2 \to L_3, \qquad \boldsymbol{B}^*: L_{3/2} \to L_2.$$

Then the functional V is defined on L_2, and the operation represented on the right-hand side of (5.10) takes L_2 into itself. (Note that in the application to periodic waves, where the functions in question are defined on a finite interval, the first condition on $\boldsymbol{B}$ and $\boldsymbol{B}^*$ is implied by the second, for then $L_3 \supset L_2 \supset L_{3/2}$.) In the existence proof for periodic solutions which is to be completed in § 5.4, this operation is shown also to be completely continuous in L_2, which implies at once that W is a weakly continuous functional (so establishing an essential element of the variational argument).

It is especially noteworthy that these variational formulations of the steady-wave problem, either as here or as explained in 2, are also potentially regarding questions of stability. There is clearly a good possibility of being able to adapt the arguments presented in § 4.

5.2. Explicit exact solutions. The KdV and BBM equations correspond to $\boldsymbol{H} = -\partial_x^2$, i.e. $\alpha(k) = k^2$, and for this example the steady solitary-wave and periodic (cnoidal) solutions are well known (as recalled in §4). The only other example for which a general class of steady solutions is known appears to be $\alpha(k) = |k|$, which was shown by Benjamin [1967a] to arise from the theory of internal waves in heterogeneous fluids of infinite depth. It is interesting that the operator $\boldsymbol{H}$ in this case may be interpreted as the positive square root, in the class of real symmetric operators, of the differential operator $-\partial_x^2$ which is its counterpart in the KdV and BBM equations. Explicit expressions for solitary-wave and periodic solutions were given in the cited paper.

It should be noted that any number of special solutions can be generated by considering the general problem in an an inverse sense. For instance, the Fourier transform of (5.1) is

$$[C - c(k)]\hat{\phi} = \frac{1}{2\pi}(\hat{\phi} * \hat{\phi}),$$

and we can specify the parameter C and function $c(k)$ to satisfy this equation when a particular $\phi(x)$ is chosen and the transform $\phi(k)$ and convolution $(\hat{\phi} * \hat{\phi})$ are evaluated. An interesting example of this kind was noted by Whitham [1967b].

5.3. Operations on periodic functions. The form of the problem applying to periodic functions now needs to be made specific. We consider real functions $v(x)$ that are periodic on $[-l, l]$, and write

$$\chi_n(x) = \exp(in\pi x/l),$$

$$\bar{\chi}_n(x) = \exp(-in\pi x/l),$$

with $n = \ldots -2, -1, 0, 1, 2, \ldots$. The Plancherel version of Fourier's expansion theorem tells us that, if $v \in L_2(-l, l)$, then

$$v(x) = \frac{1}{2l} \sum_{n=-\infty}^{\infty} a_n(v)\chi_n(x), \tag{5.11}$$

where $a_n(v) = \int_{-l}^{l} v\bar{\chi}_n \, dx = (v, \chi_n)$. We also have, by Parseval's theorem,

$$\|v\|^2 = \int_{-l}^{l} v^2 \, dx = \frac{1}{2l} \sum_{-\infty}^{\infty} |a_n(v)|^2. \tag{5.12}$$

For periodic functions, accordingly, the pseudo-differential operator $\boldsymbol{H}$ is defined by

$$(\boldsymbol{H}v, \chi_n) = \alpha(n\pi/l)a_n(v) \quad \text{if } v \in D(\boldsymbol{H}),$$

the domain of definition $D(\boldsymbol{H})$ being the subset of L_2 for which $\sum_{-\infty}^{\infty} \alpha_n|a_n|^2$ $(\alpha_n = \alpha(n\pi/l))$ converges. Similarly, the operator $\boldsymbol{B}$ introduced in § 5.1, 3(b) is now defined by

$$(B v, \chi_n) = \gamma_n a_n(v), \tag{5.13}$$

where $\gamma_n = 1/(1 + \alpha_n)^{1/2}$. As the numbers γ_n form a bounded sequence, in fact $\gamma_n \to 0$ for $|n| \to \infty$, $\boldsymbol{B}$ is thus defined on the whole of L_2.

With regard to the even positive function $\alpha(k)$, the symbol of $\boldsymbol{H}$, we assume the condition of growth with $|k|$ that was mentioned in the final paragraph of § 5.1, 2, which now takes the form

$$\gamma_n = O(|n|^{-\beta/2}) \quad \text{for } n \to \pm\infty,$$

with $\beta > 1$. This implies that $\sum \gamma_n^2$ converges; and since if $v \in L_2$ the Fourier expansion (5.11) is also an l_2 series, it follows (by the Cauchy inequality) that the expansion

$$\boldsymbol{B}v(x) = \frac{1}{2l} \sum_{-\infty}^{\infty} \gamma_n a_n \chi_n(x)$$

is an l_1 series. Hence, by the Riemann-Lebesgue theorem, $\boldsymbol{B}v$ is a continuous function. Thus the operator $\boldsymbol{B}$ has the property

$$\boldsymbol{B}: L_2 \to C, \tag{5.14}$$

and, being linear, is evidently a continuous operator in this sense.

5.4. Existence of periodic solutions. We proceed to complete a proof of the existence of nontrivial periodic solutions of (5.10). The property (5.14) of $\boldsymbol{B}$ will serve presently to establish the regularity of solutions, and it implies at once that the functional $W(v)$ given in (5.9) is defined on L_2 (although obviously, as exemplified below (5.10), weaker conditions on $\boldsymbol{B}$ would suffice for this). The crucial step in the argument is to show that $W(v)$ is *weakly continuous* on L_2, by which it is meant that if $\{v_m\}$ $(m = 1, 2, \ldots)$ is any sequence[9] in L_2 converging weakly to ψ as $m \to \infty$, then $\lim_{m\to\infty} W(v_m) = W(\psi)$.

For $W(v)$ to have this property it is sufficient that the gradient operator given by

$$\tfrac{1}{2} \operatorname{grad} W(v) = \boldsymbol{B}^*\{\boldsymbol{B}v + \tfrac{1}{2}(\boldsymbol{B}v)^2\}$$

is completely continuous (Vainberg [1964, Theorem 8.2], Berger and Berger [1968, p. 111]). In view of (5.14) the operation $\boldsymbol{B}v + \frac{1}{2}(\boldsymbol{B}v)^2$ is evidently continuous in L_2, and so we need merely show that the continuous operator $\boldsymbol{B}^* \equiv \boldsymbol{B}: L_2 \to L_2$ is completely continuous. Accordingly, from (5.13) and (5.12) we observe that

[9] For example, the sequence $\operatorname{Re}\{\chi_m\}$ whose weak limit is zero.

$$\|\boldsymbol{B}v(x+\varepsilon) - \boldsymbol{B}v(x)\|^2 = \frac{1}{2l}\sum_{-\infty}^{\infty} 2\{1 - \cos(n\pi\varepsilon/l)\}\gamma_n|a_n|^2$$

$$\leqq \frac{c}{2l}\sum_{-\infty}^{\infty} |a_n|^2 = c\|v\|^2,$$

where

$$c = \sup_n 2\{1 - \cos(n\pi\varepsilon/l)\}\gamma_n.$$

Since γ_n converges to zero as $|n| \to \infty$, the positive number c can be made arbitrarily small by choosing a small enough value of $\varepsilon > 0$. Thus the set of functions $\boldsymbol{B}v$ corresponding to $\|v\| \leqq 1$ is *equicontinuous*, and therefore relatively compact, with respect to the L_2 norm. This proves that $\boldsymbol{B}$ is a completely continuous operator, and consequently $W(v)$ is a weakly continuous functional on L_2.

Now, by virtue of its weak continuity, $W(v)$ achieves a maximum value in any ball $\|v\| \leqq R$ (Vainberg, Theorem 13.2). The point of maximum cannot be either of the points $v \equiv 0$ or $v = -2$ at which grad $W(v) \equiv 0$, not the first because obviously W can have positive values in the ball and not the second because $W(-2) < W(2)$. The maximum must therefore be achieved at a point ψ on the sphere $\|v\| = R$, and at this point the gradient of W is necessarily collinear with the normal vector (Vainberg, Theorem 12.2; Berger and Berger, p. 112), which is precisely the condition expressed by (5.10′). Thus we have established the existence of a function $\psi \in L_2$ which satisfies (5.10) and realizes the maximum of $W(v)$ for $\|v\| = V^{1/2}(v) = R$. The constant C in (5.10) is, of course, related to R and cannot be prescribed independently.

Since the constant

$$\psi_0 = R/(2l)^{1/2} = 2(C_0 - 1)$$

is trivially a solution of the variational problem (5.10′) (cf. second paragraph of § 4.5), it remains to find a specification such that $W(\psi_0)$ is not a conditional maximum.[10] To this end, we consider the second differential of $W(v)$ at $v = \psi_0$, in directions tangent to the sphere $\|v\| = V^{1/2}(\psi_0) = R$ (i.e. in directions h orthogonal to ψ_0). This is

$$D^2W(\psi_0; h) = \left[\frac{d^2}{ds^2}\{W(\psi_0 + sh) - C_0V(\psi_0 + sh)\}\right]_{s=0}$$

$$= 2\int_{-l}^{l} \{(1 + \psi_0)(\boldsymbol{B}h)^2 - C_0h^2\}\,dx,$$

[10] Note that $-\psi$ is also a solution, corresponding to a $C_0 < 1$, but it is irrelevant since $W(-\psi_0) < W(\psi_0)$.

in which the functions h are required to satisfy $\int_{-l}^{l} h\,dx = 0$. It is necessary for a maximum that $D^2 W \leq 0$ for all such h. But, taking $h = \cos(\pi x/l)$ which gives $\boldsymbol{B}h = \gamma_1 h$, we see that $D^2 W > 0$ if $(1 + \psi_0)\gamma_1^2 > C_0$, and this condition is equivalent to

$$R/(R + 2(2l)^{1/2}) > \alpha_1 = \alpha(\pi/l). \tag{5.15}$$

The function $\alpha(k)$ converges to zero as $|k| \to 0$, and we may assume that $|\alpha'(k)|$ is bounded in a neighbourhood of zero. Hence, for a given $R > 0$, the inequality (5.15) can always be satisfied by taking l sufficiently large, say $l > l_c(R)$. Thus, if $l > l_c(R)$, $W(\psi_0)$ is not a maximum for $V(v) = R^2$.

We conclude that, for any given $\|\psi\| = R > 0$, *equation* (5.10) *has a nontrivial periodic solution* ψ *over any period* $2l$ *greater than* $2l_c(R)$.

[The L_2 expansion theorem (5.11) and (5.12) may be used to show that if l is sufficiently small for the inequality opposite to (5.15) to be satisfied, then $D^2 W < 0$ for all admissible h and $W(\psi_0)$ is a conditional maximum. It seems likely that in this case ψ_0 is the only nonzero solution of (5.10).]

5.5. Properties of solutions. (i) The preceding argument establishes the solution ψ of (5.10) as an element of $L_2(-l, l)$, but the property (5.14) implies that any such solution is also a continuous function. Moreover, $\phi = \boldsymbol{B}\psi$ is a continuous function satisfying the original pseudo-differential equation (5.3). In fact, by application of a "bootstrap" argument to equation (5.10), it may now easily be deduced that ψ and ϕ are C^∞ functions.

(ii) It is obvious from the definitions in § 5.3 that if $\psi(x)$ is a solution, so is any translation $\psi(x + a)$. This is also evident from the fact that for periodic functions the functionals $V(v)$ and $W(v)$, defined by integrals over one period, are invariant under the translation group.

(iii) In view of the fact that $\boldsymbol{B}$ is a symmetric operator, the extremal property of the solution can readily be shown to imply that $\psi(x)$ is an even function if located so that $\psi'(0) = 0$. Correspondingly, $\phi(x)$ is also even.

(iv) The fact that $W(\psi)$ is an absolute maximum for $V(\psi) = R^2$ implies that $W(\psi) > W(\psi_0) = R^2(1 + \frac{1}{3}\psi_0)$ and $\int_{-l}^{l} (\boldsymbol{B}\psi)^3\,dx > 0$. (For, if this integral were negative, then $W(-\psi) > W(\psi)$.) But, after multiplying (5.10) by ψ and integrating, we have

$$CR^2 = \int_{-l}^{l} \left\{(\boldsymbol{B}\psi)^2 + \frac{1}{2}(\boldsymbol{B}\psi)^3\right\} dx = W(\psi) + \frac{1}{6}\int_{-l}^{l} (\boldsymbol{B}\psi)^3\,dx. \tag{5.16}$$

It follows that $C > (1 + \frac{1}{3}\psi_0) > 1$. Sharper lower estimates for C can be obtained by, for example, evaluating $W(v)$ for $v = a + b\cos(\pi x/l)$ with $(a^2 + \frac{1}{2}b^2)(2l) = R^2$. And by use of a test function such as $v = p\exp(-q|x|)$, with $p, q > 0$, it may be shown that $C - 1$ has a positive lower bound in the limit $l \to \infty$ for fixed R.

To obtain an upper estimate for C, we note that $\|\boldsymbol{B}\psi\| \leqq \|\psi\| = R$ and hence, from (5.16),

$$C \leqq 1 + \tfrac{1}{2}\max(\boldsymbol{B}\psi).$$

[It may easily be verified also that the solution $\phi = \boldsymbol{B}\psi$ of (5.3) has values lower as well as higher than $2(C - 1)$.] Considering the Fourier series expansion of $\boldsymbol{B}\psi$ and using the Cauchy inequality, we obtain

$$\boldsymbol{B}\psi \leqq bR, \quad \text{where } b = \left(\frac{1}{2l}\sum_{-\infty}^{\infty} \gamma_n^2\right)^{1/2},$$

so that $C \leqq 1 + \frac{1}{2}bR$.

(v) It can be shown that the maximizing solution ψ has $2l$ as its fundamental period, although other nontrivial solutions may exist having periods $2l/N$ with $N = 2, 3, \ldots$. The foregoing estimates appear to remain meaningful in the limit as $l \to \infty$ (in particular, b goes over to a convergent integral), and so it may confidently be expected that ψ becomes a solitary-wave solution in this limit. An exact proof of this property poses additional difficulties, however, and this aspect will not be covered here, although it probably affords the most general means of establishing the existence of solitary-wave solutions.

Finally, note that the variational problem can be posed meaningfully in $L_2(-\infty, \infty)$, so applying directly to the case of solitary waves. But the operator $\boldsymbol{B}$ is not then completely continuous, and thus the argument used in § 5.4 cannot be used.

References

V. I. Arnol'd 1965, *Conditions for nonlinear stability of stationary plane curvilinear flows of an ideal fluid*, Dokl. Akad. Nauk. SSSR **162**, 975–978. Soviet Math. Dokl. **6**, 773. MR **31** #4288.

——— 1966, *Sur un principe variationnel pour les écoulements stationnaires des liquides parfaits et ses applications aux problèmes de stabilité non linéaires*, J. Mécanique **5**, 29. See also: 1966, *Sur la géometrie différentielle des groups de Lie de dimension infinie et ses applications à l'hydrodynamique des fluides parfaits*, Ann. Inst. Fourier (Grenoble) **16**, fasc. 1, 319–361. MR **34** #1956.

T. B. Benjamin 1967a, *Internal waves of permanent form in fluids of great depth*, J. Fluid Mech. **29**, 559.

——— 1967b, *Instability of periodic wavetrains in nonlinear dispersive systems*, Proc. Roy. Soc. London A **299**, 59.

——— 1971, *A unified theory of conjugate flows*, Philos. Trans. Roy. Soc. London A **269**, 587.

——— 1972, *The stability of solitary waves*, Proc. Roy. Soc. London, A **328**, 153.

T. B. Benjamin, J. L. Bona and J. J. Mahony 1972, *Model equations for long waves in nonlinear dispersive systems*, Philos. Trans. Roy. Soc. A **272**, 47.

T. B. Benjamin and J. E. Feir 1967, *The disintegration of wave trains on deep water*, J. Fluid Mech. **27**, 417.

T. B. Benjamin and M. J. Lighthill 1954, *On cnoidal waves and bores*, Proc. Roy. Soc. London Ser. A **224**, 448–460. MR **17**, 911, 1437.

M. Berger and M. Berger 1968, *Perspectives in nonlinearity. An introduction to nonlinear analysis*, Benjamin, New York. MR **40** #4971.

J. L. Bona and P. J. Bryant 1973, *A mathematical model for long waves generated by wavemakers in nonlinear dispersive systems*, Proc. Cambridge Philos. Soc. **73**, 391.

R. W. Carroll 1969, *Abstract methods in partial differential equations*, Harper & Row, New York.

K. O. Friedrichs and D. H. Hyers 1954, *The existence of solitary waves*, Comm. Pure Appl. Math. **7**, 517–550. MR **16**, 413.

G. H. Keulegan and G. W. Patterson 1940, *Mathematical theory of irrotational translation waves*, J. Research Nat. Bur. Standards **24**, 47–101. MR **1**, 284.

D. J. Korteweg and G. de Vries 1895, *On the change of form of long waves advancing in a rectangular canal, and on a new type of long stationary waves*, Philos. Mag. (5) **39**, 422.

M. A. Krasnosel'skiĭ 1964, *Positive solutions of operator equations*, Fizmatgiz, Moscow, 1962; English transl., Noordhoff, Groningen. MR **26** #2862; **31** #6107.

Ju. P. Krasovskiĭ 1961, *On the theory of steady-state waves of finite amplitude*, Ž. Vyčisl. Mat. i Mat. Fiz. **1**, 836–855. (Russian) MR **25** #1731.

T. Levi-Civita 1925, *Détermination rigoureuse des ondes permanentes d'ampleur finie*, Math. Ann. **93**, 264.

H. Lamb 1932, *Hydrodynamics*, 6th ed, Cambridge Univ. Press, Cambridge.

P. D. Lax 1968, *Integrals of nonlinear equations of evolution and solitary waves*, Comm. Pure Appl. Math. **21**, 467–490. MR **38** # 3620.

J. L. Lions 1969, *Quelques méthodes de résolution des problèmes aux limites non linéaires*, Dunod; Gauthier-Villars, Paris. MR **41** #4326.

J. C. Luke 1967, *A variational principle for a fluid with a free surface*, J. Fluid Mech. **27**, 395–397. MR **35** #1269.

R. E. Meyer 1967, *Note on the undular jump*, J. Fluid Mech. **28**, 209.

D. H. Peregrine 1966, *Calculations of the development of an undular bore*, J. Fluid Mech. **25**, 321.

R. L. Seliger 1968, *A note on the breaking of waves*, Proc. Roy. Soc. London A **303**, 493.

J. J. Stoker 1957, *Water waves: The mathematical theory with applications*, Pure and Appl. Math., vol. 4, Interscience, New York. MR **21** #2438.

M. M. Vaĭnberg 1964, *Variational methods for the study of nonlinear operators*, GITTL, Moscow, 1956; English transl., Holden-Day, San Francisco, Calif. MR **19**, 567; **31** #638.

A. C. Vliegenthart 1971, *On finite-difference methods for the Korteweg-de Vries equation*, J. Engng. Math. **5**, 137.

J. V. Wehausen 1965, *Free-surface flows*, art. in *Research Frontiers in Fluid Dynamics* (Ed. Seeger and Temple), Interscience, New York.

G. B. Whitham 1967a, *Nonlinear dispersion of water waves*, J. Fluid Mech. **27**, 399–412. MR **34** #8711.

——— 1967b, *Variational methods and applications to water waves*, Proc. Roy. Soc. London A **299**, 6.

FLUID MECHANICS RESEARCH INSTITUTE, UNIVERSITY OF ESSEX

Lectures in Applied Mathematics
Volume 15, 1974

Long Waves

D. J. Benney

1. Introduction. In this article no attempt is made to give a comprehensive review of recent work on long nonlinear waves. Rather a few simple problems are examined which illustrate the mathematical features characteristic of this type of phenomenon. In particular waves in a fluid with a free surface will be of prime concern since the concepts which arise in these simple problems have their counterparts in more complicated situations.

2. Nonlinear shallow water theory. Consider the two-dimensional time dependent motion of an inviscid homogeneous fluid under the action of gravity g. Let $y = 0$ be the rigid bottom and $y = h(x, t)$ the free surface. With $\Psi(x, y, t)$ as a streamfunction, the horizontal and vertical velocity components are

$$u = \partial\Psi/\partial y, \qquad v = -\partial\Psi/\partial x. \tag{2.1}$$

The long wave parameter μ^2 is defined by

$$\mu^2 = (h_0/l)^2 \tag{2.2}$$

where h_0 is a mean depth and l is the horizontal scale of the wave. The appropriate equations of motion are

$$\Psi_{yt} + \Psi_y\Psi_{yx} - \Psi_x\Psi_{yy} = -(1/\rho)p_x, \tag{2.3}$$

$$\mu^2(-\Psi_{xt} - \Psi_y\Psi_{xx} + \Psi_x\Psi_{xy}) = -(1/\rho)p_y - g, \tag{2.4}$$

AMS (MOS) subject classifications (1970). Primary 76B15, 76–02.

subject to the boundary conditions

$$\Psi_x = 0, \qquad y = 0, \tag{2.5}$$

$$p = p_0, \qquad y = h(x, t), \tag{2.6}$$

$$h_t + \Psi_y h_x + \Psi_x = 0, \qquad y = h(x, t). \tag{2.7}$$

Here p_0 denotes constant atmospheric pressure and surface tension is neglected. A long wave analysis proceeds on the basis that μ^2 is small, but without further simplifications the problem remains nontrivial. The classical results are obtained by assuming the motion is irrotational so that with $\mu^2 = 0$ equation (2.5) is satisfied by

$$\Psi = u(x, t)y. \tag{2.8}$$

Equations (2.4) and (2.6) determine

$$p = p_0 - \rho g(y - h(x, t)), \tag{2.9}$$

while the familiar nonlinear shallow equations follow from (2.3) and (2.7), namely,

$$u_t + uu_x + gh_x = 0, \tag{2.10}$$

$$h_t + uh_x + hu_x = 0. \tag{2.11}$$

It might be noted that a constant vertical shear can be added without destroying the nature of these equations, for with

$$\Psi = \tau y^2/2 + a(x, t)y \tag{2.12}$$

we obtain

$$a_t + aa_x + gh_x = 0, \tag{2.13}$$

$$h_t + (a + \tau h)h_x + ha_x = 0. \tag{2.14}$$

It is possible to incorporate the higher order terms into equations (2.10), (2.11). For instance, corresponding to (2.8), the appropriate stream function is

$$\Psi = u(x, t)y - \mu^2 u_{xx}(x, t)\tfrac{1}{6}y^3 + O(\mu^4), \tag{2.15}$$

and upon substitution it is found that $u(x, t)$ and $h(x, t)$ satisfy the equations

$$u_t + (\tfrac{1}{2}u^2 + gh + \tfrac{1}{2}\mu^2 gh^2 h_{xx})_x = O(\mu^4), \tag{2.16}$$

$$h_t + (uh - \tfrac{1}{6}\mu^2 h^2 u_{xx})_x = O(\mu^4). \tag{2.17}$$

From (2.16) and (2.17) it is possible to specialize to small amplitudes and derive the well-known Boussinesq equations, and, for a wave in one direction, the Korteweg–de Vries equation.

3. Waves in shear flows. Here we indicate how weakly nonlinear waves propagate in the presence of vorticity. The difficulties associated with a critical layer will be avoided by confining attention to situations when there is no point in the flow where the wave and flow speeds are equal. For illustrative purposes we return to the problem posed by equations (2.3) through (2.7). The amplitude parameter ε is introduced where

$$\varepsilon = a_0/h_0, \tag{3.1}$$

a_0 being a wave amplitude and h_0 the mean depth. The stream function Ψ is then expanded in powers of ε and μ^2. In this way it is possible to find an equation governing the evolution of the free surface. However, it should be noted that this equation corresponds to a single mode and that the general initial value problem is more complicated.

The method of solution is equivalent to writing

$$\begin{aligned}\Psi(x, y, t) = \overline{\Psi}(y) + \varepsilon\Psi^{(1,0)}(x, y, t) + \varepsilon\mu^2\Psi^{(1,1)}(x, y, t) \\ + \varepsilon^2\Psi^{(2,0)}(x, y, t) + \cdots,\end{aligned} \tag{3.2}$$

and

$$\begin{aligned}p(x, y, t) = \bar{p}(y) + \varepsilon p^{(1,0)}(x, y, t) + \varepsilon\mu^2 p^{(1,1)}(x, y, t) \\ + \varepsilon^2 p^{(2,0)}(x, y, t) + \cdots,\end{aligned} \tag{3.3}$$

where $\overline{\Psi}_y = \bar{u}$ is the basic velocity profile. Upon substitution the resulting equations take the form

$$\begin{aligned}&\varepsilon\left(\Psi_{yt}^{(1,0)} + \bar{u}\Psi_{yx}^{(1,0)} - \bar{u}_y\Psi_x^{(1,0)} + \frac{1}{\rho}p_x^{(1,0)}\right) \\ &\quad + \varepsilon\mu^2\left(\Psi_{yt}^{(1,1)} + \bar{u}\Psi_{yx}^{(1,1)} - \bar{u}_y\Psi_x^{(1,1)} + \frac{1}{\rho}p_x^{(1,1)}\right) \\ &\quad + \varepsilon^2\left(\Psi_{yt}^{(2,0)} + \bar{u}\Psi_{yx}^{(2,0)} - \bar{u}_y\Psi_x^{(2,0)} + \Psi_y^{(1,0)}\Psi_{yx}^{(1,0)}\right. \\ &\qquad\qquad \left. - \Psi_x^{(1,0)}\Psi_{yy}^{(1,0)} + \frac{1}{\rho}p_x^{(2,0)}\right) + \cdots = 0,\end{aligned} \tag{3.4}$$

$$\begin{aligned}&\left(\frac{1}{\rho}\bar{p}_y + g\right) + \varepsilon\left(\frac{1}{\rho}p_y^{(1,0)}\right) + \varepsilon\mu^2\left(\frac{1}{\rho}p_y^{(1,1)} - \Psi_{xt}^{(1,0)} - \bar{u}\Psi_{xx}^{(1,0)}\right) \\ &\qquad + \varepsilon^2\left(\frac{1}{\rho}p_y^{(2,0)}\right) + \cdots = 0,\end{aligned} \tag{3.5}$$

while the boundary conditions, reduced to fixed levels, become

$$\varepsilon(\Psi_x^{(1,0)}) + \varepsilon\mu^2(\Psi_x^{(1,1)}) + \varepsilon^2(\Psi_x^{(2,0)}) + \cdots = 0, \qquad y = 0, \tag{3.6}$$

$$\begin{aligned} &\bar{p} - p_0 + \varepsilon(p^{(1,0)} + \eta\bar{p}_y) + \varepsilon\mu^2(p^{(1,1)}) \\ &\quad + \varepsilon^2\left(p^{(2,0)} + \eta p_y^{(1,0)} + \frac{\eta^2}{2}\bar{p}_{yy}\right) + \cdots = 0, \qquad y = h_0, \end{aligned} \tag{3.7}$$

$$\begin{aligned} &\varepsilon(\eta_t + \bar{u}\eta_x + \Psi_x^{(1,0)}) + \varepsilon\mu^2(\Psi_x^{(1,1)}) \\ &\quad + \varepsilon^2(\eta\eta_x\bar{u}_y + \eta_x\Psi_y^{(1,0)} + \Psi_x^{(2,0)} + \eta\Psi_{xy}^{(1,0)}) + \cdots = 0, \qquad y = h_0. \end{aligned} \tag{3.8}$$

In addition it is necessary to assume that $\eta(x, t)$ satisfies an evolution equation of the form

$$\eta_t = -c^{(0,0)}\eta_x - \mu^2 c^{(1,1)}\eta_{xxx} - \varepsilon c^{(2,0)}\eta\eta_x - \cdots \tag{3.9}$$

where the $c^{(0,0)} = c, c^{(1,1)}, c^{(2,0)}, \ldots$ are constants which are determined by a sequence of boundary value problems. Note this equation is essentially the Korteweg–de Vries equation. To the order considered here each function is separable and we may write

$$\Psi^{(1,0)} = \eta\phi^{(1,0)}(y), \qquad \Psi^{(1,1)} = \eta_{xx}\phi^{(1,1)}(y), \qquad \Psi^{(2,0)} = \eta^2\phi^{(2,0)}(y). \tag{3.10}$$

The series of resulting problems are as follows:

$$\begin{aligned} (\bar{u} - c)\phi_y^{(1,0)} - \bar{u}_y\phi^{(1,0)} &= -g, \\ \phi^{(1,0)} &= 0, \qquad y = 0, \\ (\bar{u} - c) + \phi^{(1,0)} &= 0, \qquad y = h_0. \end{aligned} \tag{3.11}$$

$$\begin{aligned} (\bar{u} - c)\phi_y^{(1,1)} - \bar{u}_y\phi^{(1,1)} &= c^{(1,1)}\phi_y^{(1,0)} - \int_{h_0}^{y} (\bar{u} - c)\phi^{(1,0)}\,dy, \\ \phi^{(1,1)} &= 0, \qquad y = 0, \\ \phi^{(1,1)} &= c^{(1,1)}, \qquad y = h_0. \end{aligned} \tag{3.12}$$

$$\begin{aligned} (\bar{u} - c)\phi_y^{(2,0)} - \bar{u}_y\phi^{(2,0)} &= \tfrac{1}{2}c^{(2,0)}\phi_y^{(1,0)} + \tfrac{1}{2}(\phi_y^{(1,0)2} - \phi^{(1,0)}\phi_{yy}^{(1,0)}), \\ \phi^{(2,0)} &= 0, \qquad y = 0, \\ \phi^{(2,0)} &= -\phi_y^{(1,0)} + \tfrac{1}{2}(c^{(2,0)} - \bar{u}_y), \qquad y = h_0. \end{aligned} \tag{3.13}$$

The linear long wave eigenfunction is

$$\phi^{(1,0)} = -g(\bar{u} - c)\int_0^y dy/(\bar{u} - c)^2, \tag{3.14}$$

and the eigenvalues c are determined by

$$\int_0^{h_0} dy/(\bar{u} - c)^2 = 1/g. \tag{3.15}$$

If $\bar{u}(y)$ is monotone then it can be shown that there are two values of c, $c_1 < u_{\min}$, $c_2 > u_{\max}$, and therefore no critical layer. In what follows we deal with one such value for c.

The functions $\phi^{(1,1)}$, $\phi^{(2,0)}$ satisfy inhomogeneous boundary value problems and for a solution to exist $c^{(1,1)}$, $c^{(2,0)}$ must take definite values. These calculations are straightforward and the results are

$$c^{(1,1)} = -\frac{1}{2}\frac{\int_0^{h_0}(\bar{u}(y)-c)^2\left(\int_0^y(\bar{u}(y')-c)^{-2}\,dy'\right)dy}{\int_0^{h_0}(\bar{u}(y)-c)^{-3}\,dy}, \tag{3.16}$$

$$c^{(2,0)} = -\frac{3g}{2}\frac{\int_0^{h_0}(\bar{u}(y)-c)^{-4}\,dy}{\int_0^{h_0}(\bar{u}(y)-c)^{-3}\,dy}. \tag{3.17}$$

Higher order terms can be calculated, and the whole procedure can be applied to other bounded flows [**1**], [**2**], [**3**]. For unbounded flows equation (3.9) is replaced by an integro-differential equation [**4**], [**5**].

4. Some nonlinear solutions. Although problems of weakly nonlinear waves are solved with relative ease many less trivial questions remain. One of these is to derive evolution equations for fully nonlinear long waves and this will be investigated here and in subsequent sections. In the notation of § 3 this corresponds to consideration of the problem with $\mu^2 = 0$ and $\varepsilon = O(1)$. From equations (2.3) through (2.7) we have to solve the equation

$$\Psi_{yt} + \Psi_y\Psi_{yx} - \Psi_x\Psi_{yy} = -(1/\rho)p_x, \tag{4.1}$$

$$0 = -(1/\rho)p_y - g, \tag{4.2}$$

subject to the boundary conditions

$$\Psi_x = 0, \qquad y = 0, \tag{4.3}$$

$$p = p_0, \qquad y = h(x,t), \tag{4.4}$$

$$h_t + \Psi_y h_x + \Psi_x = 0, \qquad y = h(x,t). \tag{4.5}$$

Since the pressure is hydrostatic $p = p_0 - \rho g(y-h)$, and it follows that (4.1) can be rewritten in the form

$$\Psi_{yt} + \Psi_y\Psi_{yx} - \Psi_x\Psi_{yy} = -gh_x. \tag{4.6}$$

A special nonlinear solution, meaningful in the linear limit, is readily found by asking for a decomposition of the form

$$\Psi = \Psi(y, h), \tag{4.7}$$

where $h(x,t)$ satisfies the equation

$$h_t = -c(h)h_x. \tag{4.8}$$

Here h represents a wave which deforms and propagates with speed $c(h)$. Upon using the assumed forms (4.7), (4.8), equations (4.6), (4.3) and (4.5) become

$$(\Psi_y - c)\Psi_{yh} - \Psi_h\Psi_{yy} = -g, \tag{4.9}$$

$$\Psi_h = 0, \qquad y = 0, \tag{4.10}$$

$$\Psi_y - c + \Psi_h = 0, \qquad y = h, \tag{4.11}$$

and the solution is

$$\Psi_h = -g(\Psi_y - c)\int_0^y dy/(\Psi_y - c)^2, \tag{4.12}$$

with the possible wave speeds $c(h)$ determined from

$$\int_0^h dy/(\Psi_y - c)^2 = 1/g. \tag{4.13}$$

Note that the above solution can be specialized to reproduce ε the perturbation expansion of § 3. This is accomplished by writing

$$h = 1 + \varepsilon\eta, \tag{4.14}$$

$$\Psi = \int \bar{u}\,dy + \varepsilon\eta\phi^{(1,0)}(y) + \varepsilon^2\eta^2\phi^{(2,0)}(y) + \cdots, \tag{4.15}$$

where

$$\eta_t = -c(h_0 + \varepsilon\eta)\eta_x = -c(h_0) - \varepsilon c'(h_0)\eta\eta_x - \tfrac{1}{2}\varepsilon^2 c''(h_0)\eta^2\eta_x - \cdots. \tag{4.16}$$

The classical nonlinear shallow water results of § 2 provide an additional check. In this case Ψ is proportional to y and we find

$$c = \pm(3(gh)^{1/2} - 2(gh_0)^{1/2}), \tag{4.17}$$

$$u = \pm 2((gh)^{1/2} - (gh_0)^{1/2}). \tag{4.18}$$

The type of analysis developed here can be applied to a variety of problems. One further example is chosen here, the propagation of long waves in a stratified shear flow. With u and v the velocity components and ρ the density, the appropriate equations for long waves are

$$\rho_t + \Psi_y\rho_x - \Psi_x\rho_y = 0, \tag{4.19}$$

$$\Psi_{yt} + \Psi_y\Psi_{yx} - \Psi_x\Psi_{yy} = -(1/\rho)p_x, \tag{4.20}$$

$$0 = -(1/\rho)p_y - g, \tag{4.21}$$

$$\Psi_x = 0, \qquad y = 0, \tag{4.22}$$

$$p = p_0, \qquad y = h, \tag{4.23}$$

$$h_t + \Psi_y h_x + \Psi_x = 0, \qquad y = h. \tag{4.24}$$

Again we seek solutions of the form

$$\rho = \rho(y, h), \qquad \Psi = \Psi(y, h), \qquad p = p(y, h), \tag{4.25}$$

where

$$h_t = -c(h)h_x. \tag{4.26}$$

Upon solving we obtain

$$p = p_0 - g \int_h^y \rho \, dy, \tag{4.27}$$

$$\rho_h = \rho_y \Psi_h/(\Psi_y - c), \tag{4.28}$$

$$\Psi_h = -(\Psi_y - c) \int_0^y p_h/\rho(\Psi_y - c)^2 \, dy, \tag{4.29}$$

with the propagation speeds c determined by

$$\int_0^h p_2/\rho(\Psi_y - c)^2 \, dy = 1. \tag{4.30}$$

5. Moment equations and conservation laws. The nonlinear long wave equations have some interesting properties in their own right and this more formal aspect is explored next. For this purpose we recast the problem in a more convenient form:

$$u_x + v_y = 0, \tag{5.1}$$

$$u_t + uu_x + vu_y = -gh_x, \tag{5.2}$$

$$v = 0, \qquad y = 0, \tag{5.3}$$

$$h_t + uh_x - v = 0, \qquad y = h. \tag{5.4}$$

When $u(x, y, 0)$, $h(x, 0)$ are prescribed, equations (5.1) through (5.4) pose an initial value problem for the unknown functions $u(x, y, t)$, $h(x, t)$. The case of persistent irrotationality corresponds to $(\partial u/\partial y)(x, y, 0) = 0$, from which it follows that $(\partial u/\partial y)(x, y, t) = 0$ and $u(x, t)$, $h(x, t)$ satisfy equations (2.10) and (2.11). Also if $u(x, y, 0) = \tau y + a(x, 0)$, with τ constant, the subsequent evolution is determined by equations (2.13) and (2.14). While these situations are very special, they do provide checks on the subsequent analysis.

The laws of conservation of mass, momentum and energy are well known for inviscid fluids. For the long wave free surface problem they are readily derived. First equation (5.4) can be rewritten in the form

$$\frac{\partial h}{\partial t} + \frac{\partial}{\partial x}\left(\int_0^h u\,dy\right) = 0, \tag{5.5}$$

where (5.1) has been used to eliminate v. Of course, this mass conservation law is true whether or not we are dealing with long waves.

Integration of (5.2) from $y = 0$ to $y = h$ leads to the momentum conservation law:

$$\frac{\partial}{\partial t}\left(\int_0^h u\,dy\right) + \frac{\partial}{\partial x}\left(\int_0^h u^2\,dy + \frac{gh^2}{2}\right) = 0, \tag{5.6}$$

where (5.1), (5.3) and (5.4) have all been used. Finally, if (5.2) is multiplied by u and integrated from $y = 0$ to $y = h$ the energy conservation law is obtained:

$$\frac{\partial}{\partial t}\left(\frac{1}{2}\int_0^h u^2\,dy + \frac{gh^2}{2}\right) + \frac{\partial}{\partial x}\left(\frac{1}{2}\int_0^h u^3\,dy + gh\int_0^h u\,dy\right) = 0. \tag{5.7}$$

The above three conservation laws are to be expected. What is unexpected is that there are an infinite number of conservation laws. The author was led to suspect the truth of this result by actually constructing the first eight of them, which are listed below. The notation is such that

$$A_n = \int_0^h u^n(x, y, t)\,dy, \tag{5.8}$$

so that $A_0 = h$, $A_1 = \int_0^h u\,dy, \ldots$, etc., and the conservation laws are

$$\partial P_n/\partial t + \partial Q_n/\partial x, \qquad n = 1, 2, \ldots, 8, \tag{5.9}$$

where it can be verified that

$$P_1 = A_0, \qquad Q_1 = A_1, \tag{5.10}$$

$$P_2 = A_1, \qquad Q_2 = A_2 + \tfrac{1}{2}gA_0^2, \tag{5.11}$$

$$P_3 = \tfrac{1}{2}A_2 + \tfrac{1}{2}gA_0^2, \qquad Q_3 = \tfrac{1}{2}A_3 + gA_0A_1, \tag{5.12}$$

$$P_4 = \tfrac{1}{3}A_3 + gA_0A_1, \qquad Q_4 = \tfrac{1}{3}A_4 + gA_0A_2 + \tfrac{1}{2}gA_1^2 + \tfrac{1}{3}g^2A_0^3, \tag{5.13}$$

$$P_5 = \tfrac{1}{4}A_4 + gA_0A_2 + \tfrac{1}{2}gA_1^2 + \tfrac{1}{2}g^2A_0^3, \qquad Q_5 = \tfrac{1}{4}A_5 + gA_0A_3 + gA_1A_2 + \tfrac{3}{2}g^2A_0^2A_1, \tag{5.14}$$

$$P_6 = \tfrac{1}{5}A_5 + gA_0A_3 + gA_1A_2 + 2g^2A_0^2A_1, \qquad Q_6 = \tfrac{1}{5}A_6 + gA_0A_4 + gA_1A_3 + \tfrac{1}{2}gA_2^2 + 2g^2A_0^2A_2 + 2g^2A_0A_1^2 + \tfrac{1}{2}g^3A_0^4, \tag{5.15}$$

$$\begin{aligned} P_7 &= \tfrac{1}{6}A_6 + gA_0A_4 + gA_1A_3 + \tfrac{1}{2}gA_2^2 + \tfrac{5}{2}g^2A_0^2A_2 + \tfrac{5}{2}g^2A_0A_1^2 + \tfrac{5}{6}g^3A_0^4, \\ Q_7 &= \tfrac{1}{6}A_7 + gA_0A_5 + gA_1A_4 + gA_2A_3 + \tfrac{5}{2}g^2A_0^2A_3 + 5g^2A_0A_1A_2 + \tfrac{5}{6}g^2A_1^3 + \tfrac{10}{3}g^3A_0^3A_1, \end{aligned} \tag{5.16}$$

$$\begin{aligned} P_8 &= \tfrac{1}{7}A_7 + gA_0A_5 + gA_1A_4 + gA_2A_3 + 3g^2A_0^2A_3 + 6g^2A_0A_1A_2 + g^2A_1^3 + 5g^3A_0^3A_1, \\ Q_8 &= \tfrac{1}{7}A_8 + gA_0A_6 + gA_1A_5 + gA_2A_4 + \tfrac{1}{2}gA_3^2 + 3g^2A_0^2A_4 + 6g^2A_0A_1A_3 + 3g^2A_0A_2^2 + 3g^2A_1^2A_2 + 5g^3A_0^3A_2 + \tfrac{15}{2}g^3A_0^2A_1^2 + g^4A_0^5. \end{aligned} \tag{5.17}$$

Each conservation law is of polynomial form (in the A_n) and the successive equations become more complicated.

In order to give a constructive proof of the existence of an infinity of conservation laws, return to (5.2), multiply by u^{n-1} and integrate from $y = 0$ to $y = h$. Upon making use of (5.1), (5.3) and (5.4) we obtain a closed set of equations for the moments:

$$\frac{\partial A_n}{\partial t} + \frac{\partial A_{n+1}}{\partial x} + ngA_{n-1}\frac{\partial A_0}{\partial x} = 0, \qquad n = 0, 1, \ldots. \tag{5.18}$$

These equations are not of conservation type as they stand nor is it immediately obvious that they can be brought into such a form. Rather than deal with the infinite set (5.18) it is desirable to introduce a generating function

$$f(x, t; z) = \sum_{n=0}^{\infty} A_n(x, t)z^n. \tag{5.19}$$

Equation (5.18) is obtained by equating powers of z^n in

$$\left(\frac{\partial}{\partial t} + \frac{1}{z}\frac{\partial}{\partial x}\right) f(x, t; z) = \left\{\frac{1}{z} - gz\frac{\partial}{\partial z}(zf(x, t; z))\right\}\frac{\partial f}{\partial x}(x, t; 0), \tag{5.20}$$

or on suppressing the x and t arguments

$$zf_t + f_x = (1 - gz^2(zf)_z)f_x(0). \tag{5.21}$$

Note the special cases considered earlier correspond to the choices

$$f = \frac{h}{1 - uz}, \qquad f = -\frac{1}{\tau z}\log\left(1 - \frac{\tau hz}{1 - az}\right). \tag{5.22}$$

Finally we prove that there exist an infinite number of conservation laws. The proof is a constructive one and is motivated by the form of (5.10) through (5.17). If we multiply (5.21) by zf, then differentiate with respect to z and multiply by gz^2, we obtain

$$\begin{aligned} &(gz^2(\tfrac{1}{2}z^2f^2)_z)_t + (gz^2(\tfrac{1}{2}zf^2)_z)_x \\ &\qquad = (gz^2(zf)_z - gz^2(gz^2(\tfrac{1}{2}z^2f^2)_z)_z)f_x(0). \end{aligned} \tag{5.23}$$

The addition of (5.21) and (5.23) yields

$$\begin{aligned} &(zf + gz^2(\tfrac{1}{2}z^2f^2)_z)_t + (f + gz^2(\tfrac{1}{2}zf^2)_z)_x \\ &\qquad = (1 - gz^2(gz^2(\tfrac{1}{2}z^2f^2)_z)_z)f_x(0). \end{aligned} \tag{5.24}$$

Note that the preceding manipulation maintains all but the last term in conservation form and the last term now involves a higher power of g and therefore z. The process can be continued. For instance at the next stage we obtain

$$\begin{aligned} &\left(zf + gz^2\left(\frac{z^2f^2}{2!}\right)_z + gz^2\left(gz^2\left(\frac{z^3f^3}{3!}\right)_z\right)_z\right)_t \\ &\qquad + \left(f + gz^2\left(\frac{zf^2}{2!}\right)_z + gz^2\left(gz^2\left(\frac{z^2f^3}{3!}\right)_z\right)_z\right)_x \\ &\qquad = \left(1 - gz^2\left(gz^2\left(gz^2\left(\frac{z^3f^3}{3!}\right)_z\right)_z\right)_z\right)f_x(0), \end{aligned} \tag{5.25}$$

and finally

$$\frac{\partial}{\partial t}\left(\sum_{n=0}^{\infty} L^n\left(\frac{z^{n+1}f^{n+1}}{(n+1)!}\right)\right) + \frac{\partial}{\partial x}\left(\sum_{n=0}^{\infty} L^n\left(\frac{z^nf^{n+1}}{(n+1)!}\right) - f(0)\right) = 0, \tag{5.26}$$

where

$$L = gz^2(\partial/\partial z). \tag{5.27}$$

In equation (5.26) each power of z yields a distinct conservation law and the proof is complete.*

6. Thin viscous layers. Finally we indicate how the long wave expansion procedure can be used to investigate waves in a viscous fluid. Here it is essential that the parameter μ times the Reynolds number be small, where μ is the long wave parameter defined earlier.

A simple problem which has been subject to much theoretical and experimental work concerns the possible wave motions on a thin liquid film in motion down an inclined plane [**6**], [7], [**8**]. Suppose the plane is inclined at an angle α $(0 < \alpha < \pi)$ to the horizontal and let g, ρ, ν and T denote gravity, fluid density, kinematic viscosity and surface tension, respectively. If x, y, z are directions down, normal to, and across the plane, the appropriate equations of motion are

$$u_x + v_y + \mu w_z = 0, \tag{6.1}$$

$$\mu(u_t + uu_x + vu_y) + \mu^2 wu_z = -(\mu/\rho)p_x + g\sin\alpha + \nu(u_{yy} + \mu^2\Delta_1 u), \tag{6.2}$$

$$\mu^2(v_t + uv_x + vv_y) + \mu^3 wv_z = -(1/\rho)p_y - g\cos\alpha + \nu\mu(v_{yy} + \mu^2\Delta_1 v), \tag{6.3}$$

$$\mu(w_t + uw_x + vw_y) + \mu^2 ww_z = -(1/\rho)p_z + \nu(w_{yy} + \mu^2\Delta_1 w), \tag{6.4}$$

where the parameter μ has been introduced to indicate the ordering procedure and $\Delta_1 = \partial^2/\partial x^2 + \partial^2/\partial z^2$ is the two-dimensional Laplace operator.

The above equations must be solved subject to the following boundary conditions: (i) zero velocity at the rigid wall $y = 0$, (ii) continuity of stress at the free surface $y = h(x, z, t)$. In mathematical form these are

$$u = v = w = 0, \qquad y = 0, \tag{6.5}$$

$$u_y = O(\mu^2), \qquad y = h, \tag{6.6}$$

$$w_y = O(\mu), \qquad y = h, \tag{6.7}$$

$$p_0 - p - T\Delta_1 h = O(\mu), \qquad y = h, \tag{6.8}$$

$$h_t + \frac{\partial}{\partial x}\int_0^h u\,dy + \mu\frac{\partial}{\partial z}\int_0^h w\,dy = 0, \qquad y = h, \tag{6.9}$$

* *Note added in proof.* Professor Miura has recently shown that there are an infinite number of local conservation laws.

where only terms necessary to find the evolution equation for h to $O(\mu)$ have been retained.

The calculation is elementary and therefore the details are omitted. Each function is expanded in powers of μ so that

$$f = \sum_{n=0}^{\infty} \mu^n f^{(n)}(x, y, z, t). \tag{6.10}$$

From (6.2), (6.5), (6.6) we solve for $u^{(0)}$ to obtain

$$u^{(0)} = -\frac{g \sin \alpha}{\nu}\left(\frac{y^2}{2} - hy\right). \tag{6.11}$$

The substitution of this result into (6.9) gives

$$h_t + \left(\frac{g \sin \alpha h^3}{3\nu}\right)_x = O(\mu). \tag{6.12}$$

A further iteration shows that

$$\begin{aligned} h_t + \left(\frac{g \sin \alpha h^3}{3\nu}\right)_x &+ \frac{\mu T}{3\rho\nu}((h^3\Delta_1 h_x)_x + (h^3 \Delta_1 h_z)_z) \\ - \mu\frac{g \cos \alpha}{12\nu}\Delta_1(h^4) &+ \mu\frac{2g^2 \sin^2 \alpha}{105\nu^3}(h^7)_{xx} = O(\mu^2), \end{aligned} \tag{6.13}$$

and the process can be continued to any order. Note that the expansion proceeds in power of μ (and not μ^2) so that potential instability phenomena are present. From this general expansion it is a simple matter to return to a weakly nonlinear theory or investigate permanent wave forms.

References

1. T. B. Benjamin, *The solitary wave on a stream with an arbitrary distribution of vorticity*, J. Fluid Mech. **12** (1962), 97–116. MR**25** #1733.

2. D. J. Benney, *Long non-linear waves in fluid flows*, J. Math. and Phys. **45** (1966), 52–63. MR**32** #6784.

3. T. B. Benjamin, J. Fluid Mech. **25** (1966), 241–270.

4. ———, J. Fluid Mech. **29** (1967), 559–592.

5. S. Leibovich, *Weakly non-linear waves in rotating fluids*, J. Fluid Mech. **42** (1970), 803–822. MR**42** #8766.

6. T. B. Benjamin, *Wave formation in laminar flow down an inclined plane*, J. Fluid Mech. **2** (1957), 554–574. MR**24** #B1050.

7. D. J. Benney, *Long waves on liquid films*, J. Math. and Phys. **45** (1966), 150–155. MR**34** #1010.

8. G. J. Roskes, J. Physics and Fluids **13** (1970), 1440–1445.

Massachusetts Institute of Technology

Lectures in Applied Mathematics
Volume 15, 1974

The Korteweg-de Vries Equation and Related Evolution Equations

Martin D. Kruskal*

Introduction. The Korteweg-de Vries [**15**] (KdV) equation

$$u_t + uu_x + u_{xxx} = 0 \tag{1.1}$$

provides a simple and useful model for describing the long-time evolution of wave phenomena in which the steepening effect of the nonlinear term uu_x is counterbalanced by dispersion. It was originally derived by Korteweg and de Vries (1895) to describe the propagation of unidirectional shallow water waves. It has many other direct physical applications; magnetohydrodynamic waves in a cold plasma (Gardner and Morikawa [**9**]), longitudinal vibrations of a harmonic discrete mass string (Zabusky [**33**], Kruskal [**16**]), ion-acoustic waves in a cold plasma (Washimi and Taniuti [**29**], pressure waves in liquid-gas bubble mixtures (van Wijngaarden [**32**]), rotating flow in a tube (Leibovich [**21**]) and longitudinal dispersive waves in elastic rods (Nariboli (**25**]).

While the many applications in themselves justify the widespread interest in this equation, its mathematical properties are of such significance as to justify the intense study of recent years. It is the first nontrivial example of a nonlinear dispersive wave system for which the solution to the initial value problem can be obtained. Many of the central properties including the exact solution are given in a series of papers by Gardner, Greene, Kruskal, Miura, et al. (G-G-K-M [**22**], [**23**], [**28**], [**18**], [**8**], also

AMS (MOS) subject classifications 1970. Primary 35-02, 35A30, 35F20, 45E99, 75B15, 75B25.

* I am deeply grateful to Alan C. Newell, whose dedicated labors to produce this article from lecture notes went far beyond the call of duty. Any errors, of course, are entirely my own responsibility.

[7]). For the KdV equation, the effect of dispersion is to prevent the formation of discontinuities. Consider (1.1) and suppose that the initial data vary slowly with the position coordinate say $\partial u/\partial x = O(\delta)$, δ small. In the initial stages, the evolution of $u(x, t)$ is determined by the first two terms in (1.1) and if the dispersive term were absent a discontinuity would form in a time given by the inverse of the maximum negative slope of the initial data $u(x, 0)$. However, as the profile of $u(x, t)$ steepens, the dispersive term assumes an increasing importance. When the slope of $u(x, t)$ becomes of order unity in some region the dispersive term takes full effect, and numerical experiments show that a fine structure develops consisting of oscillations with wavelength of order unity. Some of the fine structure propagates ahead of the discontinuity and separates into well-defined permanent waves of different amplitudes. These permanent waves are fundamental quantities of the general solution to the KdV equation and are known as solitary waves or solitons. The latter name, which suggests an analogy with particles, is appropriate since the solitary waves retain their form even after joint interaction. The solitary wave may be written

$$u(x, t) = 3c \operatorname{sech}^2 ((c/4)^{1/2}(x - ct - x_0)). \tag{1.2}$$

The wave amplitude is proportional to the wave speed, so larger solitons will eventually overtake and pass through smaller ones. The only effect of an interaction having taken place is a phase shift; namely, after the interaction, the solitons are shifted in their positions relative to where they would have been had no interaction taken place. This behavior of two solitons was verified by Zabusky [**33**] and proved by Lax [**20**]. In § 4 we will present a method by which the KdV can be solved and from which the corresponding result for N distinct solitons (where N is an arbitrary but finite number) can be derived. In § 6, we introduce a representation of KdV solitons as poles in the complex x-plane. The motion of a particular pole is influenced by the relative position of all the other poles. A means of visualizing the interaction of solitons is given by tracing the trajectories of the poles.

In general, it turns out that an initial state $u(x, 0)$ with $\int u(x, 0)\, dx > 0$ will evolve into solitons propagating to the right and a dispersing oscillatory state on the left. The asymptotic behavior of the state on the right is known; it is a finite sequence of solitons with amplitudes constant in time and increasing to the right. The long-time asymptotic behavior consists of decaying oscillations on the left and a decaying self-similar structure in the middle regions. Discussion of these regions is given in a recent paper by Ablowitz and Newell [**1**].

One of the remarkable properties of the KdV equation is that it possesses an infinite set of polynomial conservation laws. In § 3 we show how the conservation laws can be derived by the use of a generating function defined by

$$w = u - i\varepsilon w_x - \varepsilon^2 w^2/6. \tag{1.3}$$

Of interest is the result that the extremum of the nth conserved quantity, subject to the constraints that all lower order conserved quantities have prescribed values, gives a $2(n - 2)$th order differential equation, whose solution has $2(n - 2)$ free parameters which can be identified with the amplitudes (or velocities) and positions of $n - 2$ ultimately well-separated solitary waves. It would seem that there is a strong correlation between the existence of an infinite number of conservation laws and the existence of intrinsic solutions (the solitons) which interact without scattering or permanent change. There is some supporting evidence that this is also the case in other nonlinear evolution equations which have recently received much attention and which are discussed in § 2. In particular, we will give a detailed account of recent work on the sine-Gordon equation in § 7.

Also intriguing is the fact that the transformation (1.3) which generates the conservation laws also provides the key to the transformation which generates the exact solution to the KdV equation. We give a brief review of a method for solving the initial value problem and its relation with the inverse scattering problem, given by Gel'fand and Levitan **[10]**, in § 4.

Although there is a formal means for determining the exact solution, an understanding of its behavior in regions where it is highly oscillatory may be gained by use of a nonlinear WKB method. Such an approach was first suggested by Kruskal [**16**] and Whitham [**30**], [**31**] and is briefly reviewed in § 5.

In the section immediately following this introduction, we give a derivation of the Korteweg–de Vries equation as it arises in the Fermi–Pasta–Ulam problem. While the derivation is not general, it is typical of the way (1.1) is derived in all the physical contexts in which it is relevant.

2. The nonlinear one-dimensional lattice. Debye in 1914 suggested that the finiteness of the thermal conductivity of an anharmonic lattice is due to its nonlinearity. In the linear case, energy is carried unhindered by the fundamental wave propagation modes, and no diffusion equation obtains. He suggested that the modes would interact due to the nonlinearity and thereby hinder the propagation of energy. The net effect of many such nonlinear interactions (many "phonon collisions") would manifest itself in a diffusion equation with a finite transport coefficient. This suggestion motivated Fermi, Pasta and Ulam ("FPU"; their original report is reproduced in this volume) to undertake a numerical study of the one-dimensional anharmonic lattice. They argued that a smooth initial state would eventually relax to a statistical equilibrium state due to nonlinear couplings in which the energy would be equidistributed among all modes on the average. The relaxation time would then provide a measure of the diffusion coefficient.

The mathematical model used by FPU to describe their one-dimensional lattice of length L consists of a row of $N - 1$ identical masses each connected to the next (and the end ones to fixed boundaries) by nonlinear springs of length h. These springs, when compressed (or extended) by an amount Δ, exert a force

$$F = k(\Delta + \alpha\Delta^{p+1}) \tag{2.1}$$

where k is the linear spring constant, α taken positive measures the strength of nonlinearity, and p is an integer which we will take to be 1. The equations governing the dynamics of this lattice are

$$\begin{aligned} my_{itt} &= k(y_{i+1} - 2y_i + y_{i-1})(1 + \alpha(y_{i+1} - y_{i-1})), \\ &\qquad i = 1, 2, \ldots, N - 1, y_0 = y_N = 0, \end{aligned} \tag{2.2}$$

y_i being the displacement of the ith mass.

FPU usually took the energy to be in the few lowest modes of the corresponding linear problem. In the linear problem, the energy in each mode would persist unchanged forever and no new mode would be excited. In the nonlinear problem, the energy flows from the low modes to higher ones, and FPU expected this to continue until the energy becomes equidistributed over all modes accommodated in their numerical scheme. With 64 points in x-space, they had 64 different modes over which they hoped to see the energy distributed. The observed evolution could then serve as a model of thermalization for more complicated physical systems.

Now a great surprise was encountered—at least it seemed to surprise everyone who was involved in this problem or heard of it. The energy did not thermalize! In fact, when the energy was initially all in the lowest mode, after flowing back and forth among several low order modes, it eventually recollected into the lowest mode to within an accuracy of one or two percent and from there on the process approximately repeated itself.

In order to obtain an analytic handle on the process, Zabusky and Kruskal decided to treat the appropriate continuum model. Expanding the displacements y_{i+1} and y_{i-1} in terms of y_i by means of a Taylor series, one obtains the following model:

$$y_{tt} = \frac{kh^2}{m}\left(y_{xx} + \frac{h^2}{12}y_{xxxx}\right)(1 + 2\alpha h y_x). \tag{2.3}$$

We set $kh^2/m = c^2$, $2\alpha h = \varepsilon$, and $h^2/12\varepsilon = \delta^2$, and treat ε as well as h as a small parameter. The necessity of including the second approximations to the central finite difference will be seen. In fact, since FPU also used discrete time steps, a similar correction term y_{tttt} could also be included. However, in order to satisfy stability requirements (Courant–Friedrichs–Léwy for the explicit scheme used) the time step was chosen smaller than the space step. Then, since ε is small and time derivatives are equivalent to space derivatives to leading approximation, the effect of including this term will merely be to make δ^2 slightly smaller (but still positive). The equation (dropping the product of two small terms) then is

$$y_{tt} - c^2 y_{xx} = \varepsilon c^2 y_x y_{xx} + \varepsilon c^2 \delta^2 y_{xxxx}. \tag{2.4}$$

We will look for solutions which represent right-going waves. Set

$$y = y^{(0)}(\xi, T) + \varepsilon y^{(1)}(x, t) + \dots, \qquad \xi = x - ct, \quad T = \varepsilon t, \tag{2.5}$$

so that to a sufficient approximation $y(\xi, T)$ varies slowly with time in a way that will be chosen to keep the asymptotic expansion (2.5) uniformly valid. It is readily found that

$$2cy^{(0)}_{\xi T} + c^2 y^{(0)}_{\xi} y^{(0)}_{\xi\xi} + c^2\delta^2 y^{(0)}_{\xi\xi\xi\xi} = 0. \tag{2.6}$$

Writing

$$u = y^{(0)}_{\xi}, \tag{2.7}$$

$$cT/2 = \tau, \tag{2.8}$$

this condition becomes

$$u_\tau + uu_\xi + \delta^2 u_{\xi\xi\xi} = 0. \tag{2.9}$$

At this point it should be clear why it was necessary to introduce the second approximation to the finite difference $y_{i+1} - 2y_i + y_{i-1}$. If $\delta^2 = 0$, equation (2.9) has solutions which develop discontinuities in a finite time. For example, take as an initial value $u(\xi, 0) = \pi a \cos 2\pi\xi$, which corresponds to the initial conditions $y(x, 0) = a \sin 2\pi x$, $y_t(x, 0) = 0$. (Recall that, since $y_t(x, 0) = 0$, only half the amplitude goes to the right.) The maximum negative slope u_ξ decreases monotonically from $-2\pi^2 a$ at $t = 0$ to $-\infty$ at $t = 1/\pi^2 a\varepsilon c$. Thus the naive continuum approximation to (2.2) breaks down.

For δ^2 finite but small, a different picture emerges. Initially the negative slope steepens until triple derivative terms become important. At this stage the solution develops fine wiggles whose length scale is $O(\delta)$ and which first appear just to the left of the front. The existence of the fine structure suggests that a form of averaging procedure may be appropriate to remove the rapid oscillations and focus on the behavior of certain "average" macroscopic quantities. We will pursue this point in § 5 after first discussing some of the properties of this remarkable equation.

3. Conservation laws. [*Note.* In this and the following section, but not § 5, one can simplify without loss by setting $\delta = 1$.] A conservation law has the form

$$T_t + X_x = 0 \tag{3.1}$$

where T is the conserved density, and $-X$ is the flux of T. Here we seek them as functionals of $u(x, t)$, the dependent variable of the KdV equation

$$u_t + uu_x + \delta^2 u_{xxx} = 0. \tag{3.2}$$

There is a close relationship between conservation laws and constants of the motion. For example, if one assumes either that u is periodic in x or that u and its x derivatives vanish sufficiently rapidly at the (finite or infinite) ends of some interval, each conservation law (3.1) yields the constant of motion

$$I = \int T\,dx \tag{3.3}$$

where the integrals are taken over either the periodic domain or the infinite domain. The first two polynomial conservation laws for (3.2) can readily be derived and in the usual applications correspond physically to momentum and energy conservation. They are

$$u_t + (\tfrac{1}{2}u^2 + \delta^2 u_{xx})_x = 0, \tag{3.4}$$

$$(\tfrac{1}{2}u^2)_t + (\tfrac{1}{3}u^3 + \delta^2 uu_{xx} - \tfrac{1}{2}\delta^2 u_x^2)_x = 0. \tag{3.5}$$

A third and less obvious conservation law was found by Whitham [**30**],

$$(\tfrac{1}{3}u^3 - \delta^2 u_x^2)_t + (\tfrac{1}{4}u^4 + \delta^2 u^2 u_{xx} - 2\delta^2 uu_x^2 - 2\delta^4 u_x u_{xxx} + \delta^4 u_{xx}^2)_x = 0 \tag{3.6}$$

and a fourth and fifth were found by Kruskal and Zabusky [**19**]. Despite rumors that five, and later when nine were found, that nine was the maximum number available, Miura was able to find a tenth. It seemed almost certain, then, that an infinite number were available and a search was instituted for a generating functional which would supply all the polynomial conservations laws. For the KdV equation, the infinite set of conservation laws are obtainable from the equation

$$w_t + \left(\frac{1}{2}w^2 + \frac{\varepsilon^2}{18}w^3 + \delta^2 w_{xx}\right)_x = 0. \tag{3.7}$$

If

$$u = w + i\varepsilon\delta w_x + \varepsilon^2 w^2/6 \tag{3.8}$$

then

$$u_t + uu_x + \delta^2 u_{xxx} = (1 + \tfrac{1}{3}\varepsilon^2 w + i\varepsilon\delta\partial/\partial x) \cdot (w_t + (w + \tfrac{1}{6}\varepsilon^2 w^2)w_x + \delta^2 w_{xxx}) \tag{3.9}$$

and therefore if $w(x, t)$ satisfies (3.7), then $u(x, t)$ satisfies (3.2). This is an extension of the Miura transformation [**22**] suggested by Gardner. Note that (3.7) is in the form of a simple conservation law with conserved density $w(x, t)$. Solving (3.8) recursively for w formally in ascending powers of ε as a functional of u yields an infinite set of conservation laws as the coefficients. Only the even powers of ε give nontrivial conservation laws. A transformation of type (3.8) was suggested by Miura [**22**] in the first paper in the Gardner–Green–Kruskal–Miura series. He examined equations of the form

$$u_t + u^p u_x + u_{xxx} = 0, \qquad p = 1,2,3,\ldots, \tag{3.10}$$

and found that the equation with $p = 1$ could be transformed into an equation with $p = 2$ by a Ricatti-like transformation.

It can be concluded that for $p = 1,2$ there exists an infinite number of polynomial conservation laws. For $p > 2$, there are known to be exactly three polynomial conservation laws [**18**] and no evidence of any more of any form. This raises the question of the role and significance of the existence of an infinite number of conservation laws for an evolution equation. It is certainly significant that the solutions of the KdV equation consisting solely of solitons can be deduced by finding the extremum of one of the constants of motion I_n subject to constraints on all the lower constants. For example, if we look for the extremum for

$$I_3 = \int (\tfrac{1}{3}u^3 - \delta^2 u_x^2)\, dx \tag{3.11}$$

subject to the constraints that the values of $I_1 = \int u\, dx$ and $I_2 = \int \frac{1}{2}u^2\, dx$ are prescribed, we find that $u(x, t)$ must satisfy

$$0 = 2\delta^2 u_{xx} + u^2 + \mu u + \lambda \tag{3.12}$$

where μ and λ are Lagrange multipliers. We recognize (3.12) as the first integral of the stationary (in a constant velocity frame) KdV equation with μ related to the speed V(or the amplitude) and λ to the "background" value U_∞: For convenience we set the latter to zero without loss of generality by Galilean invariance. The extremizing function is unique except, of course, for translation and therefore can only evolve by steady progression, since the I_n are constants. Thus extremizing I_3 for fixed I_1 and I_2 leads to an ordinary differential equation having a one-parameter family of solutions corresponding to a solitary wave. If we extremize I_4, subject to

I_1, I_2, and I_3, constant, we obtain a fourth-order ordinary differential equation which has a two-parameter family of solutions vanishing at infinity. Each parameter can be identified with the position of one of the two waves.

We expect this concept to generalize to higher order. That is, if we extremize I_n with constraints on the lower invariants, we will obtain a $2(n-2)$th order ordinary differential equation with $n-2$ parameters representing the positions of the $n-2$ waves. A fruitful area of future investigation for nonlinear dispersive systems should involve the study of the relationship between the existence of stable solitons and the existence of conservation laws. One might expect that if a large number of conservation laws exist, then correspondingly many solitons can interact simultaneously without radiation and ultimately emerge unchanged.

There has already been some promising verification of this conjecture. Zakharov and Shabat [**34**] showed that the envelope equation

$$iu_t + u_{xx} + Ku^2u^* = 0, \qquad K > 0, \tag{3.13}$$

(derived in the context of nonlinear optics by Talanov [**35**] and Kelley [**14**], and derived independently for general nondissipative systems by Bespalov, Litvak, and Talanov [**5**], and Benney and Newell [**4**]) has an infinite number of conservation laws and has soliton solutions which except for phase changes are unaffected by joint interaction. This equation can also be reduced to an inverse scattering problem by the use of an appropriate transformation.

The "sine-Gordon" equation

$$\phi_{tt} - \phi_{xx} + \sin\phi = 0, \tag{3.14}$$

which will be discussed in § 7, has also been extensively studied, and it has been shown that "kink" solutions

$$\phi = 4\tan^{-1}\exp\frac{x - ct}{(1 - c^2)^{1/2}} \tag{3.15}$$

can interact without radiation.

The acid test of the conjecture would be to examine those equations which do not (apparently) have an infinite set of conservation laws. For example,

$$u_t + u^3u_x + u_{xxx} = 0, \tag{3.16}$$

$$\phi_{tt} - \phi_{xx} + \phi - \phi^3 = 0. \tag{3.17}$$

In such cases one might expect two solitary wave solutions to interact by radiating a certain amount of their energy in the form of dispersive waves and reappearing with smaller amplitudes and velocities. A discussion of some preliminary results is given in § 7.

4. Method of solving the KdV equation. The KdV equation was originally reduced to a linear Schrödinger equation by a transformation which had as its motivation the Ricatti transformation which related the two equations (3.10) with $p = 1$ and $p = 2$. Here we will present a derivation along similar lines but which uses (3.8) the related transformation

$$u = w + i\varepsilon\delta w_x + \varepsilon^2 w^2/6 \tag{4.1}$$

which plays a major role in deducing the infinite sequence of polynomial conservation laws. Let us now treat u as a known quantity and solve (4.1) as a Ricatti equation for w. Rewrite (4.1) in the form

$$i\varepsilon\delta w_x + \varepsilon^2(w + 3/\varepsilon^2)^2/6 = u + 3/2\varepsilon^2 \tag{4.2}$$

and use the usual transformation for linearizing the Ricatti equation

$$w + 3/\varepsilon^2 = \alpha\psi_x/\psi. \tag{4.3}$$

It is seen the choice $\alpha = 6i\delta/\varepsilon$ reduces (4.2) to the equation

$$\delta^2\psi_{xx} + \tfrac{1}{6}(u + 3/2\varepsilon^2)\psi = 0. \tag{4.4}$$

In order to obtain the form in which the Schrödinger equation (4.4) was first written, set $u = -6v$, $1/4\varepsilon^2 = \lambda$, $x = \delta X$ and $t = \delta T$.

$$\psi_{XX} - (v - \lambda)\psi = 0 \tag{4.5}$$

where $v(X, T)$ satisfies

$$v_T - 6vv_X + v_{XXX} = 0. \tag{4.6}$$

Substituting $v(X, T)$ from (4.5) into (4.6) gives

$$\lambda_T\psi^2 + [\psi Q_X - \psi_X Q]_X = 0 \tag{4.7}$$

where

$$Q \equiv \psi_T + \psi_{XXX} - 3(v + \lambda)\psi_X \tag{4.8}$$

for the time development of the solutions of equation (4.5). If ψ vanishes as $|X| \to \infty$, the second term of (4.7) vanishes on integration over the interval $(-\infty, \infty)$. Hence $\lambda_T = 0$ and the discrete eigenvalues of (4.5) are constant when $v(X, T)$ evolves according to the KdV equation (4.6).

Dropping the first term of (4.7) we can integrate twice to get

$$\psi_T + \psi_{XXX} - 3(v + \lambda)\psi_x = C\psi + D\phi. \tag{4.9}$$

Here $C(T)$ and $D(T)$ are the constants of integration and ϕ is a solution of (4.5) which is linearly independent of ψ. That is, $\phi = \int^X (dX/\psi^2)$.

It is now straightforward to deduce the evolution of ψ in regions where v vanishes, and, in particular, asymptotically for $|X| \to \infty$. For a (time-independent) discrete eigenvalue $\lambda_n < 0$, $D = 0$ because the corresponding ψ_n satisfies (4.9) and vanishes exponentially as $|X| \to \infty$ and $C = 0$ because we are assuming the normalization $\int \psi_n^2 \, dx = 1$. Then, inserting

$$\psi_n \sim c_n(T) \exp(-K_n X) \quad \text{for } X \to +\infty \tag{4.10}$$

into (4.9) with $K_n = (-\lambda_n)^{1/2} > 0$ from (4.5), we find

$$c_n(T) = c_n(0) \exp(4K_n^3 t). \tag{4.11}$$

The analogous coefficients for large negative X decay exponentially in time.

For $\lambda = k^2 > 0$, a solution of equation (4.5) for large $|X|$ is a linear combination of $\exp(\pm ikx)$. We impose on ψ the boundary conditions

$$\psi \sim \exp(-ikX) + b(k, T) \exp(ikX), \qquad X \to \infty, \tag{4.12}$$

$$\psi \sim a \exp(-ikX), \qquad X \to -\infty. \tag{4.13}$$

In the usual quantum-mechanical interpretation of equation (4.5), the coefficients of unity in (4.12) and (implied) zero in (4.13) indicate prescribed steady radiation arriving from $+\infty$ only. The coefficients of transmission $a(k, T)$ and reflection $b(k, T)$ can be shown to satisfy $|a|^2 + |b|^2 = 1$.

The spectrum for $\lambda > 0$ is continuous and we may choose λ constant, so that (4.9) is again valid. Inserting (4.12) and (4.13) into (4.9) and equating the coefficients of the two independent solutions at $+\infty$ and $-\infty$, we find $D = 0$, $C = 4ik^3$ and two equations which integrate trivially to yield

$$a(k, T) = a(k, 0), \tag{4.14}$$

$$b(k, T) = b(k, 0) \exp(8ik^3 T). \tag{4.15}$$

This information at time T is sufficient to reconstruct $v(X, T)$. We use $v(X, 0)$ to compute the reflection coefficient $b(k, 0)$ and the K_n, c_n at the initial time. Let $K(X, Y)$ for $Y \geqq X$ be the solution of the Gel'fand-Levitan equation [**10**]

$$K(X, Y) + B(X + Y) + \int_X^\infty K(X, Z) B(Y + Z) \, dZ = 0, \tag{4.16}$$

with

$$B(\xi) \equiv \frac{1}{2\pi} \int_{-\infty}^{\infty} b(k, T) \, e^{ik\xi} \, dk + \sum_n c_n^2 \exp(-K_n \xi). \tag{4.17}$$

Then

$$v(X, T) = -2 \frac{d}{dX} K(X, X). \tag{4.18}$$

The evolution of $v(X, T)$ is obtained from the explicit dependence on time of $b(k, T)$ and the c_n given by equations (4.11) and (4.15). The general solution of the Gel'fand-Levitan equation when $b(k) \not\equiv 0$ is no easy matter. When $b(k) \equiv 0$, Kay and Moses [**13**] have explicitly obtained the general solution to equation (4.16).

For practical purposes, it is often desired to determine the discrete eigenvalues of (4.5) (which Greene has referred to poetically as the soul of the general solution; the decaying oscillatory structure would then be the mortal flesh). Bounds on the eigenvalues (and the number of them) may be obtained by bounding $v(X, 0)$ everywhere from above and below by square well potentials and finding the eigenvalues of the corresponding square well potential problem. Details of this approach may be found in Schiff [**27**]. Also an upper bound on the number of solitons generated from the initial data $v(X, 0)$ may be found by a modification of Bargmann's inequality [**2**]. These techniques have been used to correlate the theoretical predictions with experimental results found by Hammack [**11**] in an experiment on water waves and are described in an article by Hammack and Segur [**12**].

The crucial property, of course, is that the eigenvalue in the transformation (4.5) is time invariant when $v(X, T)$ satisfies the KdV equation (4.6). A different argument may be presented when we are dealing with periodic functions ψ. We know from the conservation laws that if ψ is periodic

$$\frac{d}{dt}\int w\,dx = 0. \tag{4.19}$$

Integration of the original transformation (4.3) over the period gives us that $1/\varepsilon^2$ is constant. But $1/\varepsilon^2$ is directly proportional to the eigenvalue λ.

5. A nonlinear WKB method. It is seen from numerical investigations of KdV that in certain regions the solutions possess a fine oscillatory structure the length scale of which is characterized by δ. Indeed this fact is also suggested by a stationary phase analysis of $B(\xi, t)$ in (4.17) which induces a fine structure on $K(x, y, t)$ and consequently $u(x, t)$ in these regions. It would appear relevant therefore to seek a solution to

$$u_t + uu_x + \delta^2 u_{xxx} = 0, \qquad \delta^2 \ll 1, \tag{5.1}$$

which includes a fine structure of order δ. Therefore we must introduce a fast scale. However, in order to make the problem tractable we introduce the fast scale in such a way that the local structure is readily obtained. A means of accomplishing this task was proposed by Kruskal and Zabusky

[19] and Whitham [30]. The basic idea is to characterize the wiggly or fine structure as dependence on a single phase function

$$\theta = B(x, t, \delta)/\delta. \tag{5.2}$$

We define the variables L, K and l by the relations

$$L \equiv B_t, \quad K = B_x, \quad l = B_t/B_x = L/K. \tag{5.3}$$

The local structure of the solution is now described through the single phase variable θ, whereas the global or large-scale structure is recovered by treating certain macroscopic parameters (amplitude, mean and wavelength, say) as functions of the slow scales x and t.

We look for a solution of the form

$$u(x, t, \delta) = V(\theta, x, t, \delta). \tag{5.4}$$

The partial differential operators become

$$\frac{\partial}{\partial t} \to \frac{L}{\delta}\frac{\partial}{\partial\theta} + \frac{\partial}{\partial t}, \qquad \frac{\partial}{\partial x} \to \frac{K}{\delta}\frac{\partial}{\partial\theta} + \frac{\partial}{\partial x}. \tag{5.5}$$

Multiplying (5.1) by δ/K and using the new operators (5.5) yields

$$\begin{aligned} lV_\theta + VV_\theta + K^2V_{\theta\theta\theta} + \delta\left\{\frac{1}{K}(V_t + VV_x) + 3(KV)_{x\theta\theta}\right\} \\ + \delta^2\left\{\frac{1}{K}(K_{xx}V + 3(KV_x)_x)\right\}_\theta + \delta^3\frac{1}{K}V_{xxx} = 0. \end{aligned} \tag{5.6}$$

We note that, to a first approximation, the equation is an ordinary differential equation in θ which can readily be solved in terms of elliptic functions. The large-scale structure is found by imposing a periodicity on V with respect to θ and choosing the evolution of the macroscopic parameters so that the solution to the differential equation

$$lV_\theta^{(0)} + V_\theta^{(0)}V^{(0)} + K^2V_{\theta\theta\theta}^{(0)} = 0 \tag{5.7}$$

remains a uniformly valid first approximation to (5.6) for all space and time.

Introduce the asymptotic expansion

$$V = V^{(0)} + \delta V^{(1)} + \delta^2 V^{(2)} + \cdots \tag{5.8}$$

into (5.6) and equate powers of δ. As a first approximation we obtain (5.7) which when integrated twice yields

$$\tfrac{1}{2}K^2V_\theta^{(0)2} = -\tfrac{1}{6}V^{(0)3} - \tfrac{1}{2}lV^{(0)2} + mV^{(0)} + n \tag{5.9}$$

$$= \tfrac{1}{6}(\alpha - V^{(0)})(\beta - V^{(0)})(\gamma - V^{(0)}). \tag{5.10}$$

Here α, β and γ ($\alpha > \beta > \gamma$) are the three (assumed real) roots. The solution to (5.10) may be written

$$V^{(0)} = \beta + (\alpha - \beta)\mathrm{cn}^2\left(\left(\frac{\alpha - \gamma}{12}\right)^{1/2}\frac{1}{K}\theta;k\right), \qquad k = \left(\frac{\alpha - \beta}{\alpha - \gamma}\right)^{1/2}. \tag{5.11}$$

We demand that $V^{(0)}$ be of period unity in θ. This gives us the relation

$$\frac{1}{4}\left(\frac{\alpha - \gamma}{3}\right)^{1/2}\frac{1}{K} = \bar{K}(k) \tag{5.12}$$

where $\bar{K}$ is the complete elliptic function of the first kind. The parameters $\alpha, \beta, \gamma, k, K, L, l, m$ and n all depend on x and t and equation (5.12) together with the equation for the conservation of waves

$$K_t = L_x \tag{5.13}$$

holds for all x and t.

We continue to solve (5.6) iteratively. However, in order that $V^{(1)}$ be periodic of period unity, certain compatibility relations must be satisfied. There are two such conditions which include the parameters m and n. Together with (5.12) and (5.13), these equations serve to describe the evolution of the slowly varying quantities K, L, m and n. The periodicity condition (5.12) is necessary to ensure the periodicity of $V^{(1)}$. If no such condition were imposed, terms of order θ^2 would appear in $V^{(1)}$.

Another method [**30**] to derive macroscopic equations is to average conservation laws. This is by no means an equivalent method, since the nonlinear WKB method is even applicable to equations with fewer or no conservation laws, for example, equations with weak dissipation. Also, since the KdV equation possesses an infinite sequence of conservation laws, we ought to prove that the averaging method leads to only three independent macroscopic equations. This last point remains an open problem.

In the extended variables, a conservation law

$$T_t + X_x = 0 \tag{5.14}$$

has the form

$$T_t + X_x + \frac{1}{\delta}(LT + KX)_\theta = 0 \tag{5.15}$$

where T and X are polynomials in V, K and their derivatives. Integrating over a period in θ, we obtain the averaged equations

$$\langle T\rangle_t + \langle X\rangle_x = 0 \tag{5.16}$$

where $\langle F\rangle \equiv \int_0^1 F(\theta)\,d\theta$.

As we have seen, for the KdV equation the conservation laws are obtainable from

$$w_t + \left(\frac{1}{2}w^2 + \frac{\varepsilon^2}{18}w^3 + \delta^2 w_{xx}\right)_x = 0 \tag{5.17}$$

where

$$w = u - i\varepsilon\delta w_x - \tfrac{1}{6}\varepsilon^2 w^2. \tag{5.18}$$

Each conservation law is obtained by solving (5.18) recursively as for small ε, substituting in (5.17) and setting the coefficient of the appropriate power of ε to zero. Introducing a new variable $W(\theta, x, t; \delta, \varepsilon)$, the averaged equations become

$$\langle W\rangle_t + \left\langle \frac{1}{2}W^2 + \frac{\varepsilon^2}{18}W^3 + \delta^2 W_{xx}\right\rangle_x = 0 \tag{5.19}$$

where

$$W = V - i\varepsilon\delta\left(W_x + \frac{1}{\delta}KW_\theta\right) - \frac{\varepsilon^2}{6}W^2.$$

We will find that, at least to leading order (in δ), three averaged equations yield a closed system. The averaged conservation equations together with the periodicity condition (5.12) and equation (5.13) lead to a system of first order partial differential equations. Two representations of this system have been derived. One representation, due to Miura and Kruskal [**24**], is written in terms of $p = \langle U\rangle$, $q = \langle (U - p)^2\rangle$, $r = -\frac{1}{6}\langle (U - p)^3\rangle - \lambda\langle (U - p)^2\rangle$ where λ can be determined implicitly as a function of q and r. These equations are

$$p_t + pp_x + \tfrac{1}{2}q_x = 0, \tag{5.20}$$

$$q_t + 2qp_x + (p + 2\lambda + 2q\lambda_q)q_x + (5 + 2q\lambda_r)r_x = 0, \tag{5.21}$$

$$\begin{aligned} r_t + (2q\lambda + 6r)p_x &+ (-3\lambda^2 - 6(q\lambda + r)\lambda_q)q_x \\ &+ (p - 6\lambda - 6(q\lambda + r)\lambda_r)r_x = 0. \end{aligned} \tag{5.22}$$

The other representation, derived similarly by Whitham [**30**], uses as the dependent variables α, β and γ the roots of the cubic (5.10). These equations are

$$\left\{\frac{\partial}{\partial t} + \left(-l - \frac{1}{3}(\alpha - \beta)\frac{\bar{K}(k)}{\bar{K}(k) - \bar{E}(k)}\right)\frac{\partial}{\partial x}\right\}(\beta + \gamma) = 0, \tag{5.23}$$

$$\left\{\frac{\partial}{\partial t} + \left(-l - \frac{1}{3}(\alpha - \beta)\frac{(1 - k)\bar{K}(k)}{\bar{E}(k) - (1 - k)\bar{K}(k)}\right)\frac{\partial}{\partial x}\right\}(\gamma + \alpha) = 0, \tag{5.24}$$

$$\left\{\frac{\partial}{\partial t}+\left(-l+\frac{1}{3}(\alpha-\beta)\frac{(1-k)\bar{K}(k)}{k\bar{E}(k)}\right)\frac{\partial}{\partial x}\right\}(\alpha+\beta)=0, \tag{5.25}$$

where $k^2 = (\alpha - \beta)/(\alpha - \gamma)$, $-l = (\alpha + \beta + \gamma)/3$ and $\bar{K}(k)$ and $\bar{E}(k)$ are the complete elliptic integrals of the first and second kind respectively.

For a more detailed discussion of these systems of equations we refer the reader to the papers of Miura and Kruskal [**24**] and Whitham [**30**].

6. Polar representations of solitary waves. One of the remarkable properties of the solitary waves is their relative insensitivity to interaction. Indeed after interaction only a phase shift indicates that any interaction has taken place at all. During the interaction itself, however, the processes are rather complicated. One might hope, therefore, to find a representation of solitons in which more details of the interaction process could be brought to light.

Consider a single solitary wave

$$u = 3c\,\mathrm{sech}^2\,\tfrac{1}{2}c^{1/2}(x - ct). \tag{6.1}$$

If we consider x as a complex variable, then the function u may be considered as a meromorphic function with double poles at the positions

$$x - ct = s\pi i/c^{1/2}, \qquad s \text{ odd}. \tag{6.2}$$

Anticipating that it might be easier to think in terms of simple poles, the transformation

$$u = 2v_x \tag{6.3}$$

is suggested. The equation which $v(x, t)$ satisfies may be taken as

$$v_t + v_x^2 + v_{xxx} = 0 \tag{6.4}$$

for which the solitary wave solutions are given by

$$v(x, t) = 3c^{1/2}\tanh\tfrac{1}{2}c^{1/2}(x - ct). \tag{6.5}$$

The hyperbolic tangent can be expressed as (principal part) the sum of the leading terms of its Laurent expansion around each pole

$$\tanh z = \sum_{\text{all odd } s} 1/(z - \tfrac{1}{2}s\pi i) \tag{6.6}$$

and therefore

$$v(x, t) = 6\sum_{\text{all odd } s} 1/(x - ct - s\pi i/c^{1/2}). \tag{6.7}$$

One may think, therefore, of the motion of a soliton as a "parade of poles." At $t = 0$ the solitary wave is represented by a sequence of poles at the positions $x = s\pi i/c^{1/2}$, s odd, and as time evolves, the poles move at uniform velocity c. We note that each pole has strength 6 and that the faster a soliton the denser its sequence of poles.

To find the motion of a pole in general (not just when it is part of an isolated soliton) we return to equation (6.4) and seek solutions

$$v(x, t) = \frac{a(t)}{x - \xi(t)} + w(x, t) \tag{6.8}$$

where $w(x, t)$ is an analytic function at $x = \xi$. Substituting in (6.4), we find that

$$a_t = 0, \qquad a = 6, \tag{6.9}$$

and

$$w_t + (w_x)^2 + w_{xxx} + 6(\xi_t - 2w_x)/(x - \xi)^2 = 0. \tag{6.10}$$

In order that $w(x, t)$ be regular at $x = \xi$ we must therefore also have the two further conditions

$$\xi_t = 2w_x|_{x=\xi(t)} \tag{6.11}$$

and

$$w_{xx}|_{x=\xi} = 0. \tag{6.12}$$

The expression (6.11) then gives the (complex) velocity of the pole. A little calculation shows that (6.12) is a subsidiary condition in the sense that if it is satisfied at any one time, it remains satisfied for all time by virtue of (6.10).

For example, consider the row of poles (6.7) representing the single solitary wave $v = 3c^{1/2} \tanh \frac{1}{2}c^{1/2}(x - ct)$. In particular, we verify the speed of the pole,

$$\xi = r\pi i/c^{1/2}, \qquad r \text{ odd}.$$

Then

$$w(x, t) = 6 \sum_{\text{all odd } s \neq r} \frac{1}{x - ct - s\pi i/c^{1/2}} \tag{6.13}$$

and

$$2w_x|_{x=ct+r\pi i/c^{1/2}} = \frac{12c}{\pi^2}\left(\sum_{\text{all odd } s \neq r} \frac{1}{(s - r)^2}\right) = c. \tag{6.14}$$

Thus we confirm that all poles "in" a single soliton move at the same speed c. It may also be readily verified that

$$w_{xx}|_{x=ct+r\pi i/c^{1/2}} = 0. \tag{6.15}$$

The interaction of two solitons can be viewed similarly. Consider

$$(6.16)\quad v(x,t) = \sum_{\text{all odd } s} \frac{6}{x - x^{(1)} - \xi_s^{(1)}} + \sum_{\text{all odd } s} \frac{6}{x - x^{(2)} - \xi_s^{(2)}}.$$

If the solitons are initially far apart, the poles at $t = 0$ are in the positions

$$(6.17)\qquad x = x^{(1)} + s\pi i/c_1^{1/2}, \qquad s \text{ odd},$$

and

$$(6.18)\qquad x = x^{(2)} + s\pi i/c_2^{1/2}, \qquad s \text{ odd}.$$

The motion of each pole is given by (6.11), each pole being influenced by all the other poles. If the rows of poles are initially far apart, the pole $x = x^{(1)} + \xi_r^{(1)}$ will mainly feel the effects of the poles which belong to the same soliton and so

$$\xi_r^{(1)} \sim r\pi i/(c^{(1)})^{1/2} + c^{(1)}t.$$

If $c^{(1)} > c^{(2)}$ and $x^{(1)} < x^{(2)}$ then the $c^{(1)}$ soliton will move faster and eventually the poles of both solitons will influence each other. To take a convenient though special example, let $c^{(1)} = 16$, $c^{(2)} = 4$; then the poles of soliton 1 are twice as dense as the poles of soliton 2 (see Figure 1).

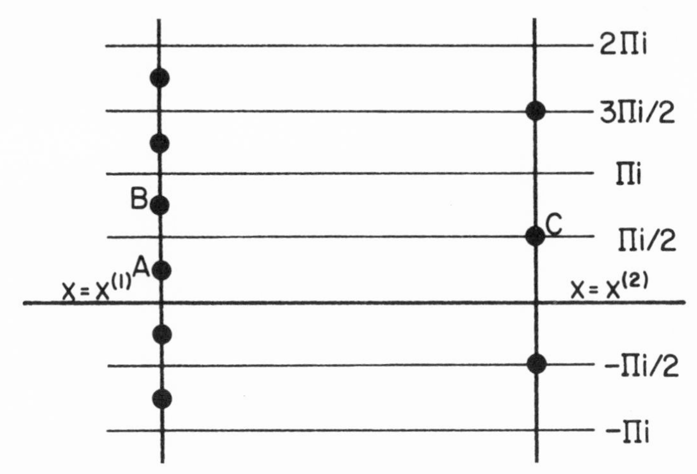

FIGURE 1. Initial pole locations for two soliton with amplitude ratio 4:1.

The pattern has periodicity πi as well as symmetry about the real axis. Initially (i.e., for large negative t) the poles A, B, C move on the paths

$$x = x^{(1)} + \pi i/4 + 16t, \quad x = x^{(1)} + 3\pi i/4 + 16t, \quad x = x^{(2)} + \pi i/2 + 4t.$$

As A and B approach C, their imaginary parts approach $\pi i/2$. The three poles eventually coalesce equi-angularly and emerge following paths which eventually return the poles to their original spacings with the larger, faster soliton on the right.

If two solitons of nearly the same amplitude interact, say $c^{(1)} = 49$, $c^{(2)} = 36$, then we know that the smaller one grows as the forward tail of the larger one affects its rear tail, and ultimately becomes the forward projection of the larger. At the same time, the larger shrinks to become the smaller. In the pole representation, this effect is achieved by some of the poles of the larger soliton (those whose imaginary parts are farthest from those of the smaller) accelerating into the gaps between the poles of the front soliton. These poles then become part of the front soliton, which because of the denser alignment of poles, now moves faster and ahead of the rear one. The pole sequences re-adjust to have equal pole spacings of $2\pi i/7$ in the front one and $2\pi i/6$ in the rear one.

The trajectories of the poles are traced by (6.11) together with the compatibility condition (6.12). Alternatively a compatible set of pole locations are any that satisfy (6.12), each pole making the second derivative of the sum of contributions of all the others vanish. This gives one equation per pole but no temporal information. Any such set of pole locations is an allowed configuration (symmetric with respect to the real axis if one wishes $v(x, t)$ to be real) and evolves into similar sets by (6.11) with no contribution arising other than the pure pole functions of strength 6.

An alternative means of finding the trajectories is suggested by a different, more convenient, form of solution of the KdV equation when only solitary waves are present. This formulation was first suggested by Hirota and has been extensively used by Whitham (unpublished). The transformation

$$u = 12\frac{\partial^2}{\partial x^2}\log F \tag{6.19}$$

yields a convenient representation for the solutions for KdV when the initial conditions contain only solitary waves. A single soliton solution is then represented by

$$F = 1 + f_1, \qquad f_1 = \exp(2k_1(x + x_1) - 8k_1^3 t) \tag{6.20}$$

with speed $4k_1^2$. Two solitons are represented by

$$F = 1 + f_1 + f_2 + \left(\frac{k_1 - k_2}{k_1 + k_2}\right)^2 f_1 f_2 \tag{6.21}$$

with speeds $4k_1^2$ and $4k_2^2$. But since $u = 2v_x$,

$$v = 6\frac{d}{dx}\log F + K, \qquad K \text{ a constant}, \tag{6.22}$$

and thus the poles of $v(x, t)$ are given by the zeros of $F(x, t)$. But $F(x, t)$ is known explicitly and thus the trajectories of the zeros can be followed precisely.

7. The sine-Gordon equation. In 1939, Frenkel' and Kontorova [**6**] studied the propagation of a "slip" in an infinite chain of elastically bound atoms lying over a fixed lower chain of similar atoms. To describe this effect they obtained a difference-differential equation which can be approximated as a partial differential equation, a nonlinear analog of the Klein-Gordon equation which has been called the "sine-Gordon equation" (Rubinstein [**26**]),

$$\phi_{tt} - \phi_{xx} + \sin\phi = 0. \tag{7.1}$$

The equation also describes the mechanical system consisting of a torsion bar horizontal in a gravitational field and free to turn on a fixed central axis to which a series of weights is attached [**3**]. Here ϕ is the angle between the downward direction and the direction from the axis to the weight and x is the distance along the central axis. The family of localized steady progressive wave solutions of (7.1) of the form

$$\phi = 4\tan^{-1}\exp((x - ct - x_0)/(1 - c^2)^{1/2}), \tag{7.2}$$

called "kinks," have the expected property of Lorentz invariance. This solution has a twist of one revolution in the chain of weights. There is also a family of "antikinks" with reverse twist. Just as the solitary wave

$$u = 3c\,\mathrm{sech}^2\,(c/4)^{1/2}(x - ct)$$

plays a central role in the solutions of the KdV equation, it is expected that the kinks and antikinks are fundamental entities for the sine-Gordon equation.

It is readily verified that the more general equation

$$\phi_{tt} - \phi_{xx} + V'(\phi) = 0 \tag{7.3}$$

has the conservation laws

$$(\phi_x\phi_t)_t - (\tfrac{1}{2}\phi_t^2 + \tfrac{1}{2}\phi_x^2 - V)_x = 0, \tag{7.4}$$

$$(\tfrac{1}{2}\phi_x^2 + \tfrac{1}{2}\phi_t^2 + V)_t - (\phi_x\phi_t)_x = 0, \tag{7.5}$$

expressing conservation of momentum and energy, as well as the trivial but meaningful law $(\phi_x)_t - (\phi_t)_x = 0$ expressing conservation of twist. With some algebra, it can be shown that for a further conservation law involving up to second derivatives of ϕ to exist, it is necessary and sufficient that $V''' = kV'$ for some constant k, one solution of which is $V' = \sin \phi$.

We can demonstrate that (7.1) possesses an infinite sequence of conservation laws. (The method was found by deliberate analogy with the KdV case.) It is convenient to rotate the axes. Let

$$(x - t)/2 = \xi, \tag{7.6}$$

$$(x + t)/2 = \eta, \tag{7.7}$$

whereupon we find

$$\phi_{\xi\eta} = \sin \phi. \tag{7.8}$$

Upon direct substitution of the transformation

$$\phi = \psi - \sin^{-1} \varepsilon\psi_\xi \tag{7.9}$$

we find

$$\phi_{\xi\eta} - \sin \phi = \left((1 - \varepsilon^2\psi_\xi^2)^{1/2} - \varepsilon\frac{\partial}{\partial\xi}\right)\left(\frac{\psi_{\xi\eta}}{(1 - \varepsilon^2\psi_\xi^2)^{1/2}} - \sin\psi\right). \tag{7.10}$$

Therefore if ψ satisfies

$$\psi_{\xi\eta}/(1 - \varepsilon^2\psi_\xi^2)^{1/2} = \sin\psi, \tag{7.11}$$

then ϕ satisfies (7.8). (I am indebted to David Wiley for the elegance of this specific form of this transformation and some related contributions.) A simple conservation law for (7.11) may be found by multiplying by ψ_ξ. (There are actually infinitely many, as is easily seen a posteriori.) We find

$$\frac{\partial}{\partial\eta}\left(\frac{(1 - \varepsilon^2\psi_\xi^2)^{1/2} - 1}{\varepsilon^2}\right) - \frac{\partial}{\partial\xi}\cos\psi = 0. \tag{7.12}$$

Expanding (7.9) in powers of ε, solving recursively for ψ in terms of ϕ, and substituting into (7.12) yields an infinite sequence of conservation laws for the sine-Gordon equation. Only the even powers of ε give nontrivial laws. To each such law there is also a dual law obtained by interchanging ξ and η, since (7.8) is invariant under that interchange. In particular from $O(\varepsilon^2)$ the dual laws (7.4) and (7.5) are obtained.

Just as for KdV, the transformation which generates the infinite sequence of polynomial conservation laws also generates the transformation by which the sine-Gordon equation may be linearized. If we set

$$\gamma = \tan \tfrac{1}{2}(\psi - \phi) \tag{7.13}$$

then

$$\gamma_\xi = \tfrac{1}{2}(1 + \gamma^2)(\psi_\xi - \phi_\xi). \tag{7.14}$$

But

$$\psi_\xi = \frac{1}{\varepsilon}\sin(\psi - \phi) = \frac{1}{\varepsilon}\frac{2\gamma}{1 + \gamma^2},$$

and therefore

$$\gamma_\xi = \frac{\gamma}{\varepsilon} - \frac{1+\gamma^2}{2}\phi_\xi. \tag{7.15}$$

The linearizing transformation

$$\gamma - \frac{1}{\varepsilon\phi_\xi} = \frac{2}{\phi_\xi}\frac{\chi_\xi}{\chi} \tag{7.16}$$

is immediately suggested, whereupon we obtain

$$\left(\frac{1}{\phi_\xi}\chi_\xi\right)_\xi - \left(\frac{1}{4\varepsilon^2\phi_\xi} + \frac{1}{2\varepsilon}\frac{\phi_{\xi\xi}}{\phi_\xi^2} - \frac{\phi_\xi}{4}\right)\chi = 0 \tag{7.17}$$

analogous to the Schrödinger problem for KdV. It remains to be shown that the "eigenvalue" $1/\varepsilon$ for (7.17) with some appropriate boundary conditions remains invariant as ϕ evolves according to (7.8).

As a final remark, it should be noted that there is a striking similarity between the solitons for the KdV equation and the kink solutions of the sine-Gordon equation. Indeed, both of these special solutions have the property that they preserve their identity throughout a nonlinear interaction, the only memory being a phase shift. Recently, I have investigated other nonlinear Klein-Gordon equations which admit kink-like solutions. The results were obtained by numerical integration of the equation

$$\phi_{tt} - \phi_{xx} + F(\phi) = 0. \tag{7.18}$$

(This work was done in collaboration with M. J. Ablowitz.) The two particular force functions $F(\phi)$ considered were

$$F_1(\phi) \begin{array}{ll} = \pi/4, & 0 < \phi < \pi, \\ = -\pi/4, & \pi < \phi < 2\pi, \end{array} \qquad F_1(\phi + 2\pi) = F_1(\phi), \tag{7.19}$$

$$F_2(\phi) = -\phi + \phi^3/\pi^2. \tag{7.20}$$

In both cases a kink-like steady progressive wave solution $\phi = \Phi(x - ct)$ exists which monotonically crosses from one stable rest value of v to another (e.g., $0 \to 2\pi$ in (7.19), $-\pi \to \pi$ in (7.20)). However, when such a wave is made to interact with its corresponding anti-wave (which crosses back from $2\pi \to 0$ or $\pi \to -\pi$), in neither case do they emerge from the interaction unaffected. In particular, the amount of radiation emitted from the interaction depends on the relative speed at which the waves approach each other. When their relative velocities are large, the waves are slowed down but maintain their structure. However, when their relative velocity is small enough, they interact for a long time ("capture" each other) and radiate their energy.

These results suggest, but certainly do not prove, that the nondistortion (after interaction) of solitons and kinks is intimately related to the existence of infinitely many conservation laws for the KdV and sine–Gordon equations respectively.

References

1. M. J. Ablowitz and A. C. Newell, J. Mathematical Phys. **14** (1973), 1277.

2. V. Bargmann, *On the number of bound states in a central field of force*, Proc. Nat. Acad. Sci. U.S.A. **38** (1952), 961–966. MR **14**, 875.

3. A. Barone, F. Esposito, C. J. Magee and A. C. Scott, Rev. Nuovo Cimento **1** (1971), 227.

4. D. J. Benney and A. C. Newell, *The propagation of nonlinear wave envelopes*, J. Mathematical Phys. **46** (1967), 133–139. MR **39** #2397.

5. V. I. Bespalov, A. G. Litvak and V. I. Talanov, Second All-Union Sympos. on Nonlinear Optics Collection of Papers, "Nauka", Moscow, 1968 (Russian).

6. Ja. I. Frenkel' and T. Kontorova, *On the theory of plastic deformation and twinning*, Acad. Sci. USSR J. Phys. **1** (1937), 137–149. MR **1**, 190.

7. C. S. Gardner, J. M. Greene, M. D. Kruskal and R. M. Miura, Phys. Rev. Lett. **19** (1967), 1095.

8. ———, (to appear).

9. C. S. Gardner and G. K. Morikawa, Sciences Report #NYO-9082, 1960 (unpublished).

10. I. M. Gel'fand and B. M. Levitan, *On the determination of a differential equation from its spectral function*, Izv. Akad. Nauk SSSR Ser. Mat. **15** (1951), 309–360; English transl., Amer. Math. Soc. Transl. (2) **1** (1955), 253–304. MR **13**, 558; **17**, 489.

11. J. Hammack, Ph.D. Thesis, Caltech., 1972.

12. J. Hammack and H. Segur, (to appear).

13. I. Kay and H. E. Moses, *The determination of the scattering potential from the spectral measure function.* III. *Calculation of the scattering potential from the scattering operator for the one-dimensional Schrödinger equation*, Nuovo Cimento (10) **3** (1956), 276–304. MR **17**, 971.

14. P. L. Kelley, Phys. Rev. Lett. **15** (1965), 1005.

15. D. J. Korteweg and G. de Vries, Philos. Mag. **39** (1895), 422.

16. M. D. Kruskal, *Asymptotology*, Proc. Conf. on Math. Models in the Physical Sciences (Univ. of Notre Dame, South Bend, Ind., 1962), Prentice-Hall, Englewood Cliffs, N.J., 1963, p. 17.

17. ———, Proc. IBM Scientific Computing Sympos. on Large-Scale Problems in Physics, IBM Data Processing Division, White Plains, New York; Thomas J. Watson Research Center, Yorktown Heights, N.Y., 1965, p. 43.

18. M. D. Kruskal, R. M. Miura, C. S. Gardner and N. J. Zabusky, *Korteweg-de Vries equation and generalizations.* V. *Uniqueness and nonexistence of polynomial conservation laws*, J. Mathematical Phys. **11** (1970), 952–960. MR **42** #6410.

19. M. D. Kruskal and N. J. Zabusky, Princeton Plasma Physics Laboratory Annual Report MATT-Q-21, 1963, p. 301 (unpublished).

20. P. D. Lax, *Integrals of nonlinear equations of evolution and solitary waves*, Comm. Pure Appl. Math. **21** (1968), 467–490. MR **38** #3620.

21. S. Leibovich, *Weakly non-linear waves in rotating fluids*, J. Fluid Mech. **42** (1970), 803–822. MR **42** #8766.

22. R. M. Miura, *Korteweg-de Vries equation and generalizations.* I. *A remarkable explicit nonlinear transformation*, J. Mathematical Phys. **9** (1968), 1202–1204. MR **40** #6042a.

23. R. M. Miura, C. S. Gardner and M. D. Kruskal, *Korteweg-de Vries equation and generalizations.* II. *Existence of conservation laws and constants of motion.* J. Mathematical Phys. **9** (1968) 1204–1209.

24. R. M. Miura and M. D. Kruskal, SIAM J. Appl. Math. (to appear).

25. G. A. Nariboli, 1969 (unpublished).

26. J. Rubinstein, *Sine-Gordon equation*, J. Mathematical Phys. **11** (1970), 258–266. MR **41** #4958.

27. L. I. Schiff, *Quantum mechanics*, McGraw-Hill, New York, 1949.

28. C. H. Su and C. S. Gardner, *Korteweg-de Vries equation and generalizations.* III. *Derivation of the Korteweg-de Vries equation and Burgers equation*, J. Mathematical Phys. **10** (1969), 536–539. MR **42** #6409.

29. H. Washimi and T. Taniuti, Phys. Rev. Lett. **17** (1966), 996.

30. G. B. Whitham, *Non-linear dispersive waves*, Proc. Roy. Soc. Ser. A **283** (1965), 238–261. MR **31** #996.

31. ———, *A general approach to linear and non-linear dispersive waves using a Lagrangian*, J. Fluid Mech. **22** (1965), 273–283. MR **31** #6459.

32. L. van Wijngaarden, J. Fluid Mech. **33** (1968), 465.

33. N. J. Zabusky, Proc. Sympos. on Nonlinear Partial Differential Equations (Univ. of Delaware, Newark, Del., 1965), Academic Press, New York, 1967, p. 223.

34. V. E. Zakharov and A. B. Shabat, JETP **34** (1972), 62–69.

35. V. I. Talanov, JETP Lett. **2** (1967), 141.

PRINCETON UNIVERSITY

Lectures in Applied Mathematics
Volume 15, 1974

Periodic Solutions of the KdV Equations

Peter D. Lax

In this article we describe some recent results on nonlinear equations of evolution in general and the Korteweg–de Vries (KdV) equation in particular. Specifically we shall study the long-term behavior of solutions and departure from ergodicity.

Before turning to nonlinear equations, a brief outline of the facts of life concerning conservative linear equations, i.e. equations of the form

$$u_t = Au, \tag{1}$$

where A is antisymmetric

$$A^* = -A. \tag{2}$$

For solutions of such an equation, energy, defined as (u, u), is conserved, where the parentheses denote the scalar product in the underlying Hilbert space. More generally, *any quadratic quantity Q of the form*

$$Q(u) = (u, Bu) \tag{3}$$

is conserved, provided that the operator B commutes with A. This is easily verified by differentiation:

$$\begin{aligned} Q_t = (u, Bu)_t &= (u_t, Bu) + (u, Bu_t) \\ &= (Au, Bu) + (u, BAu) = (u, [BA - AB]u), \end{aligned}$$

which is zero if and only if B commutes with A.

AMS (*MOS*) *subject classifications* (1970). Primary 34–00, 35–00.
Key words and phrases. Korteweg–de Vries equation, Hill's equation.

Another well-known fact is that *if B commutes with A, its nullspace is an invariant subspace for solutions of* (1). For clearly, $v = Bu$ satisfies

$$v_t = Bu_t = BAu = ABu = Av;$$

it follows now from conservation of energy that if v is initially zero, it is zero for all t.

The complete story about equations of the form (1) is contained in the spectral resolution of A:

$$A = \int \lambda \, dE(\lambda),$$

$E(\lambda)$ a projection valued measure. Solutions of (1) can be written as

$$u(t) = \int e^{i\lambda t} \, dE(\lambda) f, \qquad f = u(0). \tag{4}$$

If A has a pure point spectrum, the measure E is discrete and the integral (4) is a sum:

$$u(t) = \sum e^{i\lambda_k t} a_k e_k, \tag{4'}$$

where e_k are eigenvectors of A, $i\lambda_k$ the corresponding eigenvalues, and a_k the Fourier coefficients of the initial value f. Formula (4′) shows that *if the spectrum of A is discrete, solutions of* (1) *are almost periodic functions of t.* On the other hand, suppose the spectrum of A is absolutely continuous, which means that for any two vectors f and g, the measure $(E(\lambda)f, g)$ is absolutely continuous with respect to Lebesgue measure. Then

$$(u(t), g) = \int e^{i\lambda t} \, d(E(\lambda)f, g) \tag{4''}$$

is the Fourier transform of an L_1 function, and so according to the Riemann-Lebesgue lemma tends to zero as t tends to ∞. Thus (4″) shows that *if the spectrum of A is absolutely continuous, solutions of* (1) *decay weakly to zero as t tends to ∞.*

We turn now to nonlinear equations

$$u_t = K(u). \tag{5}$$

We assume that the initial value problem is properly posed, both in the forward and backward direction, for all initial values in a given linear space; for simplicity we take this space of initial data to be a Hilbert space. $K(u)$ is some nonlinear operator, defined on a subset of the underlying Hilbert space.

For nonlinear equations there is no simple recipe like (3) for constructing conserved quantities: there may be none. In what follows we hypothesize the existence of a conserved functional $J(u)$; of course there is no reason why J should be quadratic.

Whenever there is a functional J conserved for solutions of (5), the set $J(u) = \text{const}$ is an invariant set for solutions of (5). This set, having codimension 1, is rather large; we shall show now how to construct with the aid of J a much smaller one, analogous to the nullspace of B. We observe that B is the Fréchet derivative of the quadratic conserved quantity Q in (3); suppose the functional J is Fréchet differentiable, with Fréchet derivative $G(u)$:

$$\frac{d}{d\varepsilon}J(u(\varepsilon)) = \left(G(u), \frac{du}{d\varepsilon}\right). \tag{6}$$

THEOREM. *Let $J(u)$ be a conserved functional for solutions of* (5), *Fréchet differentiable, with Fréchet derivative $G(u)$. Then the nullset of $G(u)$ is an invariant subset for solutions of* (5).

PROOF. Let n be any vector in the nullset of G:

$$G(n) = 0. \tag{7}$$

Denote by $u(t)$ that solution of (5) whose initial value is n. Denote by $u(t, \varepsilon)$ any one-parameter family of solutions of (5) which at $\varepsilon = 0$ reduce to $u(t)$, and which depend differentiably on ε. For example, we can take $u(\varepsilon, t)$ to be the solution of the following initial value problem:

$$u(T, \varepsilon) = u(T) + \varepsilon w, \tag{8}$$

where T is an arbitrary time and w an arbitrary vector. We assume that there is a dense set of w for which the dependence of u on ε is differentiable.

Since J is a conserved quantity for all solutions of (5), $J(u(t, \varepsilon))$ is independent of t for all ε. So then is its derivative with respect to ε at $\varepsilon = 0$; according to (6), this is

$$(G(u(t)), v(t)) \tag{9}$$

where

$$v(t) = \left.\frac{du(t, \varepsilon)}{d\varepsilon}\right|_{\varepsilon=0}. \tag{10}$$

By definition $u(0) = n$, and, by assumption (7), $G(n) = 0$; so it follows that (9) is zero at $t = 0$; since (9) is independent of t, it is zero for all t, in particular $t = T$:

$$(G(u(T)), v(T)) = 0. \tag{11}$$

It follows from (8) and (10) that $v(T) = w$; since w can be chosen arbitrarily in a dense set, it follows from (11) that $G(u(T)) = 0$. Since T is arbitrary, the proof of the theorem is complete.

The usefulness of this theorem lies in reducing the study of solutions to a smaller set. This method is applicable to equations which possess Fréchet differentiable conserved functionals, whose derivatives G possess a nontrivial and yet tractable nullspace. These conditions are met by the Korteweg–de Vries equation, abbreviated as KdV:

$$u_t + uu_x + u_{xxx} = 0. \tag{12}$$

We shall study two classes of solutions:

(i) periodic solutions,

(ii) solutions defined on the whole x-axis which vanish together with all their derivatives at $x = \pm\infty$.

The KdV equation is blessed with many conserved quantities; the following three are classical:

$$J_0(u) = \int u\,dx, \qquad J_1(u) = \int \tfrac{1}{2}u^2\,dx, \qquad J_2(u) = \int (\tfrac{1}{3}u^3 - u_x^2)\,dx. \tag{13}$$

The integration is over a single period in case (i), over the whole real axis in case (ii).

Zabusky and Kruskal [**Z–K**] and Miura, Gardner and Kruskal [**M–G–K**] have discovered that (13) is merely the first three of an infinity sequence of conserved quantities. The next one is

$$J_3 = \int (\tfrac{1}{4}u^4 - 3uu_x^2 + \tfrac{9}{5}u_{xx}^2)\,dx. \tag{13'}$$

All the integrals of this series are integrals of polynomials of u and its derivatives; therefore the Fréchet derivatives of linear combinations of them are ordinary nonlinear differential operators; in $[\mathbf{L}]_2$ we have investigated the nullspace of linear combinations of the gradients of J_1, J_2 and J_3 in case (ii); it would also be interesting to study these nullspaces in the periodic case.*

We turn now to another kind of conserved functionals for KdV, discovered by Gardner, Greene, Kruskal and Miura [**G–G–K–M**]; these

---

* *Added in proof.* The author, with the help of Mac Hyman, has recently completed a theoretical and computer study of solutions of KdV whose initial values satisfy

$$J_3(u) + AJ_2(u) + BJ_1(u) = 0$$

and are periodic in x. These solutions turn out to be periodic in time as well except for a phase shift.

are the eigenvalues of the Schrödinger operator

$$L = D^2 + \tfrac{1}{6}u. \tag{14}$$

The best way to see this is to ask how the potential $u/6$ may be deformed through a one-parameter family without changing the eigenvalues of L. Under such a deformation the operators $L(t)$ remain unitarily equivalent, i.e.,

$$U^*(t)L(t)U(t) = L(0) \tag{15}$$

for some one-parameter family of unitary operators $U(t)$. Let us assume that $U(t)$ is differentiable in t; then

$$U_t = BU, \tag{16}$$

where $B(t)$ is antisymmetric

$$B^* = -B. \tag{17}$$

Differentiating (15) and using (16) and (17) we get

$$L_t = BL - LB = [B, L]. \tag{18}$$

$L_t = u_t$ is a scalar operator; it was observed in $[\mathbf{L}]_2$ that (18) has an infinite sequence of solutions $B_0, B_1, B_2, \ldots$, where B_q is a linear antisymmetric differential operator of order $2q + 1$. The first of these, $B_0 = D$, leads to a trivial result but the next

$$B_1 = -4D^3 - uD - \tfrac{1}{2}u_x, \tag{19}$$

satisfies $[B_1, L] = -\frac{1}{6}u_{xxx} - \frac{1}{6}uu_x$; when it is substituted into (18) we are led to the KdV equation, thus proving the contention that the eigenvalues of L do not change if u changes pursuant to the KdV equation. For B_q of the above series (18) yields

$$u_t = K_q(u), \tag{20}$$

where K_q is a nonlinear differential operator of order $2q + 1$. We shall call the qth equation the KdV equation of order q; for all of them, the eigenvalues of L are conserved functionals.

Relation (15) can be rewritten as

$$L(t)U(t) = U(t)L(0). \tag{21}$$

Let ϕ_0 be an eigenfunction of $L(0)$:

$$L(0)\phi_0 = \lambda\phi_0.$$

Applying (21) to ϕ_0 we get

$$L(t)U(t)\phi_0 = \lambda U(t)\phi_0$$

which shows that $\phi(t) = U(t)\phi_0$ is an eigenfunction of $L(t)$. Differentiating this relation with respect to t and using the differential equation (16) satisfied by U we get

$$\phi_t = B\phi. \tag{22}$$

For B of form (19),

$$\phi_t + 4\phi_{xxx} + u\phi_x + \tfrac{1}{2}u_x\phi = 0. \tag{23}$$

This third order equation can be turned into a first order equation by recalling that ϕ, being an eigenfunction of L, satisfies

$$\phi_{xx} + \tfrac{1}{6}u\phi = \lambda\phi.$$

Differentiating with respect to x and substituting into (23) we get

$$\phi_t + (4\lambda + \tfrac{1}{3}u)\phi_x - \tfrac{1}{6}u_x\phi = 0. \tag{23'}$$

In case (ii) when the interval is infinite, L has only a finite number of eigenvalues, all of which are positive; the rest of the spectrum is continuous, filling up the negative axis, with multiplicity 2. To each point $-k^2$ of the continuous spectrum one can assign two linearly independent generalized eigenfunctions, solutions of

$$\phi_{xx} + \tfrac{1}{6}u\phi = -k^2\phi. \tag{24}$$

Since u tends to zero rapidly as $|x|$ tends to ∞, it follows from (24) that for $|x|$ large, ϕ is very nearly a linear combination of the exponentials e^{ikx} and e^{-ikx}. The solutions ψ_+ and ψ_- which are singled out have the following behavior at $\pm\infty$:

$$(25)_l \qquad \begin{aligned} \psi_l(x) &\simeq e^{ikx} + R_l\, e^{-ikx} && \text{near } -\infty, \\ &\simeq T_l e^{ikx} && \text{near } +\infty, \end{aligned}$$

and

$$(25)_r \qquad \begin{aligned} \psi_r(x) &\simeq T_r\, e^{-ikx} && \text{near } -\infty, \\ &\simeq e^{-ikx} + R_r\, e^{ikx} && \text{near } +\infty. \end{aligned}$$

The quantities $T_l = T_l(k)$ and $R_l = R_l(k)$ are called transmission and reflection coefficients from the left, T_r and R_r from the right. Suppose $\phi(x, 0)$ is normalized as in $(25)_l$; since for large x both u and u_x are very small, equation (23′) becomes, with $\lambda = -k^2$,

$$\phi_t - 4k^2\phi_x = 0.$$

Since, for large $|x|$, ϕ is a linear combination of e^{ikx} and e^{-ikx}, it follows that

$$\begin{aligned} \phi(x, t) &\simeq e^{4ik^3t}\, e^{ikx} + R_l\, e^{-4ik^3t}\, e^{-ikx} && \text{near } -\infty, \\ &\simeq Te^{4ik^3t}\, e^{ikx} && \text{near } +\infty. \end{aligned}$$

To bring ϕ into the form $(25)_l$ we have to multiply ϕ by e^{-4ik^3t}; this reveals that

$$(26)_l \qquad T_l(k, t) = T_l(k, 0), \qquad R_l(k, t) = e^{-8ik^3t}R_l(k, 0).$$

Similarly

$$(26)_r \qquad T_r(k, t) = T_r(k, 0), \qquad R_r(k, t) = e^{8ik^3t}R_r(k, 0).$$

These relations were derived by Gardner, Greene, Kruskal and Miura [**G–G–K–M**] who observed furthermore that the reflection and transmission coefficients, together with the point eigenvalues (as well as the so-called normalizing factors of the eigenfunctions), constitute a complete set of variables from which the potential $\frac{1}{6}u$ can be uniquely reconstructed. The logarithm of the reflection coefficient plays the role of angle variable. Indeed, Faddeev and Zakharov [**F–Z**] have shown that on the whole real axis the KdV equation can be regarded as a completely integrable Hamiltonian system; another Hamiltonian treatment is due to Gardner [**G**].

It is not known whether the KdV equation is completely integrable in the periodic case. No analogue of an angle variable is known in this case; some light can be shed on this question by studying the behavior of solutions on the invariant subsets described at the beginning of this article.

The concluding part of this article is about some fragmentary results which can be deduced from equation (23′). That equation is most naturally studied in terms of its characteristics:

$$(27) \qquad dx/dt = 4\lambda + \tfrac{1}{3}u.$$

Along these characteristics (23′) becomes an ordinary differential equation:

$$(28) \qquad d\phi/dt = \tfrac{1}{6}u_x\phi.$$

In particular, we conclude that *the zeros of $\phi(x, t)$ propagate along characteristics.*

In a film depicting the time evolution of an initially sinusoidal solution of the KdV equation, made by Gary Deem at Bell Laboratories under the direction of Kruskal and Zabusky, there are eight clearly distinguishable "solitons", each pursuing its own path. It is tempting to try to relate the paths of these solutions to the characteristics (27).

Suppose in the interval $[x_1, x_2]$, $\phi(x, 0)$ is free of zeros. Then, since zeros of ϕ follow the characteristics, the whole strip between the characteristics $x_1(t)$ and $x_2(t)$ issuing from these points will be free of zeros of $\phi(x, t)$. The following peculiar conservation law holds:

THEOREM. *If ϕ is an eigenfunction of $L = D^2 + \frac{1}{6}u$, where u is a solution of the KdV equation, and $x_1(t)$ and $x_2(t)$ are two characteristics satisfying* (27), *then*

$$\Phi = \int_{x_1(t)}^{x_2(t)} \frac{1}{\phi^2(x, t)} dx \tag{29}$$

is independent of t.

PROOF. Differentiate and use equation (23′) to express ϕ_t; we get

$$\Phi_t = \int_{x_1}^{x_2} -\frac{2\phi_t}{\phi^3} dx + \frac{dx}{dt}\frac{1}{\phi^2}\Bigg|_{x^1}^{x_2}$$

$$= \int_{x_1}^{x_2} \left[\left(4\lambda + \frac{u}{3}\right)\frac{2\phi_x}{\phi^3} - \frac{1}{3}\frac{u_x}{\phi^2}\right] dx + \frac{dx}{dt}\frac{1}{\phi^2}\Bigg|_{x_1}^{x_2}.$$

The integrand is the derivative of $-(4\lambda - \frac{1}{3}u)/\phi^2$. So we deduce that

$$\Phi_t = -\frac{4\lambda + \frac{1}{3}u}{\phi^2}\Bigg|_{x_1}^{x_2} + \frac{1}{\phi^2}\frac{dx}{dt}\Bigg|_{x_1}^{x_2}.$$

Formula (27) for dx/dt shows that Φ_t is indeed 0.

Kruskal and Zabusky have observed recurrence[1] for solutions of the KdV equation; that is, there exist times T when $u(x, T)$ returns very close to $u(x, 0)$. The conservation of the quantity (29) might be of some help in explaining this phenomenon. Suppose there is a time T when all characteristic curves of (27) return to their initial position:

$$x(T) = x(0).$$

It follows then from the constancy of Φ that $\phi(x, T)$ also must be equal to $\phi(x, 0)$. If the return of the characteristics to their initial position after time T were only approximate, we could still conclude that $\phi(x, T)$ is approximately equal to $\phi(x, 0)$. Furthermore if there were a time T when all characteristics corresponding to all the important eigenvalues of L returned approximately to their initial position, then at time T the corresponding eigenfunctions would be close to their initial values. This would imply that $\frac{1}{6}u(x, t)$, the potential at time T, would be close to $\frac{1}{6}u(x, 0)$, the potential at time $t = 0$. However I do not see at this time any direct way of showing the simultaneous recurrence of all characteristics corresponding to all significant eigenfunctions.

[1] Such recurrence of the notation of a nonlinearly coupled chain was originally observed by Fermi, Pasta and Ulam.

We now specialize these considerations to so-called cnoidal solutions of KdV which are of the form

$$u(x, t) = s(x - ct), \tag{30}$$

i.e. waves traveling with speed c and without altering their form. Substituting this wave form into (12) we get

$$-cs' + ss' + s''' = 0.$$

Integrating once gives

$$-cs + \tfrac{1}{2}s^2 + s'' = \tfrac{1}{2}a$$

where a is a constant of integration. Multiplying by $2s'$ we get after integration

$$s'^2 = -\tfrac{1}{3}s^3 + cs^2 + as + b; \tag{31}$$

this equation has periodic solutions which are elliptic functions.

If the dependence of u on t is as in (30), if follows that the eigenfunctions of $L = D^2 + \frac{1}{6}u$ depend similarly on t:

$$\phi(x, t) = \phi(x - ct). \tag{32}$$

We saw earlier that the zeros of the eigenfunctions propagate along characteristics, i.e. they satisfy equation (27):

$$dx/dt = 4\lambda + \tfrac{1}{3}u(x, t).$$

Since the zeros of (32) propagate along $x = ct + x_0$, it follows that

$$\frac{dx}{dt} = c = 4\lambda + \frac{s(x - ct)}{3} = 4\lambda + \frac{s(ct + x_0 - ct)}{3},$$

i.e. that

$$c = 4\lambda + \tfrac{1}{3}s(x_0). \tag{33}$$

For large values of λ equation (33) cannot be satisfied! Thus we conclude that eigenfunctions of $L = D^2 + s/6$ corresponding to λ large enough have no zeros at all. This contradicts the classical oscillation theorem about differential equations, according to which the higher eigenfunctions have more and more zeros! The only way out of that paradox[2] is that all these higher values of λ are double eigenvalues of L, so that the proper form of equation (32) is

$$\phi(x, t) = a_1(t)\phi_1(x - ct) + a_2(t)\phi_2(x - ct).$$

---

[2] I acknowledge with pleasure a useful conversation with Jürgen Moser on this subject.

The significance of all but a finite number of eigenvalues of an operator K being double lies in the stability of the Hill equation associated with L:

$$D^2 y + \tfrac{1}{6} s y + \lambda y = 0. \tag{34}$$

According to a classical theory, see Coddington and Levinson [**C**–**L**] or Magnus and Winkler [**M**–**W**], one has to consider the sequence of eigenvalues of L under the boundary condition of periodicity with *twice* the period of s:

$$\mu_0 < \mu_1 \leqq \mu_2 < \mu_3 \leqq \mu_4 < \cdots. \tag{35}$$

Note that half the values (35) namely $\mu_0, \mu_3, \mu_4, \mu_7, \mu_8, \cdots$ correspond to eigenfunctions whose period is the same as that of s. Solutions of (34) remain bounded for all x if λ lies in one of the *intervals of stability* (μ_0, μ_1), (μ_2, μ_3), (μ_4, μ_5), and solutions are unbounded in the *intervals of instability*

$$(\mu_1, \mu_2), (\mu_3, \mu_4), \cdots. \tag{36}$$

If, as in our case, all but a finite number of eigenvalues are double, it follows that *all but a finite number of the instability intervals* (36) *disappear.*

We can estimate quite precisely how many are left. The periodic solutions defined by (31) have exactly one maximum and one minimum over a single period, and so two maxima and minima over a double period. It follows that there are at most four values of x_0 that can satisfy equation (33), and therefore eigenfunctions with more than four zeros belong to multiple eigenvalues. This shows that all eigenvalues from μ_5 on are double. We claim that already $\mu_3 = \mu_4$; the eigenfunctions corresponding to these eigenvalues have the same period as s. If μ_3 were $< \mu_4$, then the zeros x_0 satisfying (33):

$$c = 4\mu_3 + \tfrac{1}{3} s(x_0)$$

and those satisfying

$$c = 4\mu_4 + \tfrac{1}{3} s(x_0)$$

would not separate each other, contrary to the separation theorem for differential equations. So we conclude that *the Hill equation* (34) *has only one instability interval.*

Hochstadt, in his interesting paper [**H**], has shown that, conversely, if a Hill equation $D^2 y + qy + \lambda y = 0$ with a periodic potential q has only one instability interval, the potential q is an elliptic function.

We have indicated earlier that there is a whole sequence of generalized KdV equations (20) which have the property that if u is subject to (20), the operators $L(t) = D^2 + \frac{1}{6}u$ remain unitarily equivalent:

$$L(t)U(t) = U(t)L(0).$$

It follows as before that the eigenfunctions ϕ satisfy

$$\phi_t = B_q\phi, \tag{37}$$

where $B_q = U_tU^*$ is an operator of order $2q + 1$, with leading term const D^{2q+1}, whose coefficients depend on u and its derivatives. Equation (37) is of order $2q + 1$; it can be reduced to an equation of first order with the aid of the eigenvalue relation $L\phi = \lambda\phi$. The zeros of ϕ propagate along the characteristics of this first order equation, i.e. they satisfy an equation of the form

$$dx/dt = P(\lambda, u) \tag{38}$$

where P is a polynomial in λ whose highest coefficient is constant, and the rest of the coefficients are functions of u and its derivatives.

Suppose u is a traveling wave, i.e. of the form

$$u(x, t) = s_q(x - ct),$$

c constant, $s_q(x)$ a periodic function. If λ is a simple eigenvalue, we deduce as before that $\phi(x, t) = \phi(x - ct)$ and therefore the zeros of $\phi(x, t)$ satisfy $x = ct + x_0$. It follows then from (38) that $c = P(\lambda, s_q(x_0))$. For λ large enough this equation is not satisfied by any x_0; since the higher eigenfunctions of L do have zeros, it follows that all eigenvalues λ greater than some number are double. According to the theory of the Hill equation just described, it follows that *the Hill equation*

$$D^2y + \tfrac{1}{6}s_qy + \lambda y = 0$$

has only a finite number of instability intervals. Here s_q is a periodic solution of

$$cs_q + K_q(s_q) = 0;$$

K_q is a generalized KdV operator.

It would be interesting to ascertain if the converse of this proposition is true.†

We refer to Chapter VII of Magnus and Winkler for the literature on this subject.

References

[C–L] E. A. Coddington and N. Levinson, *Theory of ordinary differential equations*, McGraw-Hill, New York, 1955. MR**16**, 1022.

---

†*Added in proof.* Hochstadt has informed me recently that a student of his has verified this conjecture for two instability intervals.

[**F–Z**] L. Faddeev and L. D. Zakharov, *Korteweg–de Vries equation as completely integrable Hamiltonian system*, Funkcional. Anal. i Priložen. **5** (1971), 18–27. (Russian)

[**G**] C. S. Gardner, *Korteweg–de Vries equation and generalizations*. IV. *The Korteweg–de Vries equation as a Hamiltonian system*, J. Mathematical Phys. **12** (1971), 1548–1551. MR**44** #3615.

[**G–G–K–M**] C. S. Gardner, J. M. Greene, M. D. Kruskal and R. M. Miura, *A method for solving the Korteweg–de Vries equation*, Phys. Rev. Lett. **19** (1967), 1095–1097.

[**G–M**] C. S. Gardner and G. K. Morikawa, *Similarity in the asymptotic behavior of collision-free hydromagnetic waves and water waves*, New York Univ., Courant Inst. Math. Sci. Report NY0-9082, 1960.

[**H**] H. Hochstadt, *On the characterization of a Hill equation via its spectrum*, Arch. Rational Mech. Anal. **19** (1965), 353–362. MR**31** #6019.

[**KdV**] D. J. Korteweg and G. de Vries, *On the change of form of long waves advancing in a rectangular channel and on a new type of long stationary wave*, Philos. Mag. **39** (1895), 422–443.

[**L**]$_1$ P. D. Lax, *Nonlinear partial differential equations of evolution*, Proc. Internat. Congress Math. (Nice, 1970), Gauthier-Villars, Paris, 1971, pp. 831–840.

[**L**]$_2$ ———, *Integrals of nonlinear equations of evolution and solitary waves*, Comm. Pure Appl. Math. **21** (1968), 467–490. MR**38** #3620.

[**M–W**] W. Magnus and S. Winkler, *Hill's equation*, Interscience, Tracts in Pure and Appl. Math., no. 20, Wiley, New York, 1966. MR**33** #5991.

[**M**] R. M. Miura, *Korteweg–de Vries equation and generalizations*. I. *A remarkable explicit nonlinear transformation*, J. Mathematical Phys. **9** (1968), 1202–1204. MR**40** #6042a.

[**M–G–K**] R. M. Miura, C. S. Gardner and M. D. Kruskal, *Korteweg–de Vries equations and generalizations*. II. *Existence of conservation laws and constants of motion*, J. Mathematical Phys. **9** (1968), 1204–1209. MR**40** #6042b.

[**S**] A. Sjoberg, *On the Korteweg–de Vries equation*, J. Math. Anal. Appl. **29** (1970), 569–579.

[**W**] G. B. Whitham, *Nonlinear dispersives waves*, Proc. Roy. Soc. Ser. A. **283** (1965), 238–261. MR**31** #996.

[**Z**] N. J. Zabusky, *A synergetic approach to problems of nonlinear dispersive wave propagation and interaction*, Nonlinear Partial Differential Equations, Academic Press, New York, 1967.

[**Z–K**] N. J. Zabusky and M. D. Kruskal, *Interaction of solutions in a collisionless plasma and the recurrence of initial states*, Phys. Rev. Lett. **15** (1965), 240–243.

COURANT INSTITUTE OF MATHEMATICAL SCIENCES, NEW YORK UNIVERSITY

Lectures in Applied Mathematics
Volume 15, 1974

Two-Timing, Variational Principles and Waves[1]

G. B. Whitham

In this paper, it is shown how the author's general theory of slowly varying wave trains may be derived as the first term in a formal perturbation expansion. In its most effective form, the perturbation procedure is applied directly to the governing variational principle and an averaged variational principle is established directly. This novel use of a perturbation method may have value outside the class of wave problems considered here. Various useful manipulations of the average Lagrangian are shown to be similar to the transformations leading to Hamilton's equations in mechanics. The methods developed here for waves may also be used on the older problems of adiabatic invariants in mechanics, and they provide a different treatment; the typical problem of central orbits is included in the examples.

1. Introduction. The purpose of this paper is to show how the variational theory for slowly varying nonlinear wave trains (Whitham [**1965a**], [**1965b**], [**1967a**], [**1967b**]) may be formally justified as the first term in a consistent perturbation expansion. The theory referred to uses an "averaged" variational principle, and the derivation of the equations governing the slow variations of a nearly periodic wave train is then surprisingly general and simple. The averaging procedure is intuitively

AMS (*MOS*) *subject classifications* (1970). Primary 76B15, 35D99.

[1] From *Two-timing, variational principles and waves* by Gerry Whitham which appeared in the Journal of Fluid Mechanics, **44** (1970), 373–395. Reprinted by permission of Cambridge University Press.

This research was supported by the Office of Naval Research, U.S. Navy.

correct, but it is not immediately clear how the averaged variational principle would appear as the first term of a detailed expansion. Luke [**1966**] has shown how the final results can be justified in such a scheme, but he works with the Euler equations rather than directly on the variational principle. This has the disadvantages that the simplicity and conciseness of the variational technique are lost and the formal justification no longer follows the intuitive derivation. It does, however, justify the results and has its own points of interest; it can be extended to dissipative systems, for example, where a variational principle may not be known.

It has now been seen how to apply the "two-timing" approach used by Luke directly on the variational principle. This is the main topic of the paper. A second topic concerns some subtleties in the actual use of the variational principle. Various equivalent forms of the averaged variational principle can be introduced, and one of them has enormous advantages over the others. In specific problems, the optimum one can usually be found without too much difficulty. However, there are subtleties involved which were not clearly recognized in the original versions. These are now explained, and general procedures are given for finding the preferred form.

The original theory is reviewed briefly in the next section, and the perturbation schemes are then taken up in §§ 3 to 6. The questions regarding manipulations of the form of the averaged variational principle are discussed in § 7, with illustrative examples in § 8.

The justification of the averaged variational principle in §§ 5 and 6 is the result of joint discussions with Mr. M. E. Delaney.

It is a great pleasure to contribute to this volume, and in this way acknowledge a long-standing and continuing debt to Sydney Goldstein. Both as teacher and friend, his help and encouragement have always been invaluable.

2. The averaged variational principle. In all problems where the equations admit uniform periodic wave trains as solutions it appears to be generally true that the system can be derived from a variational principle. The simplest case occurs when there is one dependent variable $u(\mathbf{x}, t)$ and the variational principle takes the form

$$\delta \int\int L(u_t, u_{\mathbf{x}}, u)\, dt\, d\mathbf{x} = 0. \tag{1}$$

The Euler equation is

$$\frac{\partial L_1}{\partial t} + \frac{\partial L_2}{\partial \mathbf{x}} - L_3 = 0, \tag{2}$$

where

$$L_1 = \frac{\partial L}{\partial u_t}, \qquad L_2 = \frac{\partial L}{\partial u_{\boldsymbol{x}}}, \qquad L_3 = \frac{\partial L}{\partial u}. \tag{3}$$

If $\boldsymbol{x}$ is a vector with components x_i, the quantities $u_{\boldsymbol{x}}$ and L_2 are vectors with components $\partial u/\partial x_i$, $\partial L/\partial u_{x_i}$, respectively, and the second term in (2) is the divergence

$$\frac{\partial L_2}{\partial \boldsymbol{x}} = \frac{\partial}{\partial x_i}\left(\frac{\partial L}{\partial u_{x_i}}\right).$$

We are concerned with dispersive wave problems in which the equation (2) for u has special solutions

$$u = U_0(\theta), \qquad \theta = \kappa_i x_i - \omega t,$$

where $\boldsymbol{\kappa}$, ω are constants and where $U_0(\theta)$ turns out to be a periodic function of θ. Then $\boldsymbol{\kappa}$, ω are the wave-number and frequency, respectively. Since (2) is a second-order equation in u, there will be two constants of integration. One will determine the amplitude, a; the other will be an arbitrary shift in the phase θ and may be omitted here. The three parameters ω, $\boldsymbol{\kappa}$, a will not be independent, but must satisfy a "dispersion relation"

$$G(\omega, \boldsymbol{\kappa}, a) = 0, \tag{4}$$

if (2) is to be satisfied. In linear problems

$$U_0 = a \cos \theta, \tag{5}$$

and the dispersion relation (4) does not involve the amplitude a.

For slowly varying wave trains, it is argued that the form of the solution, $u = U_0(\theta, a)$ is maintained, but a will not be constant nor will θ be linear in $\boldsymbol{x}$ and t. The wave-number $\boldsymbol{\kappa}$ and frequency ω are generalized by defining them as

$$\omega = -\theta_t, \qquad \kappa_i = \theta_{x_i}. \tag{6}$$

The parameters ω, $\boldsymbol{\kappa}$, a will now be slowly varying functions of $\boldsymbol{x}$ and t corresponding to the slow modulation of the wave train. The aim is to derive equations for them.

If the period of the function $u = U_0(\theta, a)$ is normalized to 2π, the averaged Lagrangian is defined to be

$$\mathscr{L}(\omega, \boldsymbol{\kappa}, a) = \frac{1}{2\pi}\int_0^{2\pi} L \, d\theta, \tag{7}$$

and is calculated by substituting the *uniform* periodic solution $u = U_0(\theta, a)$

in L. The dependence on ω, $\boldsymbol{\kappa}$ comes from introducing $u_t = -\omega U_0'$, $u_x = \boldsymbol{\kappa} U_0'$ into $\mathscr{L}$; the integration with respect to θ is carried out holding ω, $\boldsymbol{\kappa}$, a constant. The claim is, then, that the "averaged" variational principle holds and that the equations for $(\omega, \boldsymbol{\kappa}, a)$ follow from

$$\delta \iint \mathscr{L}(\omega, \boldsymbol{\kappa}, a)\, dt\, d\boldsymbol{x} = 0. \tag{8}$$

The quantities ω and $\boldsymbol{\kappa}$ are defined by (6) in terms of the phase function θ, so we must derive the consequences of (8) for independent variations $\delta\theta$ and δa. From the variation δa we have

$$\mathscr{L}_a(\omega, \boldsymbol{\kappa}, a) = 0, \tag{9}$$

and from the variation $\delta\theta$ we have

$$\frac{\partial}{\partial t} \mathscr{L}_\omega - \frac{\partial}{\partial x_i} \mathscr{L}_{\kappa_i} = 0. \tag{10}$$

Equation (9) is a functional relation between $(\omega, \boldsymbol{\kappa}, a)$, and it can only be the dispersion relation (4). One can use (10) as a second-order equation in θ, or work with ω, $\boldsymbol{\kappa}$, and supplement (10) by the consistency equations

$$\frac{\partial \kappa_i}{\partial t} + \frac{\partial \omega}{\partial x_i} = 0, \qquad \frac{\partial \kappa_i}{\partial x_j} - \frac{\partial \kappa_j}{\partial x_i} = 0, \tag{11}$$

derived from (6). The latter course is usually preferable.

The study of the consequences of (9)–(11), the interpretation of $\mathscr{L}_\omega$, $\mathscr{L}_{\kappa_i}$ as "adiabatic invariants", the use of these equations to extend the concept of group velocity to nonlinear problems, the relation of the stability of the periodic wave with the type of the equations, as well as extensions of the theory to cases with more than one dependent variable and to nonhomogeneous media, are all given in the previous papers.

Two special cases will be useful in the present discussion. First, in any linear problem, L will be quadratic in u. Hence, when $U_0 = a \cos \theta$ is substituted in (7), $\mathscr{L}$ *must* take the form

$$\mathscr{L}(\omega, \boldsymbol{\kappa}, a) = F(\omega, \boldsymbol{\kappa}) a^2. \tag{12}$$

Quite generally, for linear problems with higher-order derivatives of u or with more dependent variables u, the same argument goes through. The Lagrangian L must be quadratic in the variables u, and, as a consequence, $\mathscr{L}$ must be quadratic in a and take the form given in (12). Then the dispersion relation (9) becomes

$$F(\omega, \boldsymbol{\kappa}) = 0. \tag{13}$$

Hence, without detailed calculation, the function $F(\omega, \boldsymbol{\kappa})$ in (12) *must* be the dispersion function for the problem. We also see in general that the stationary value for $\mathscr{L}$ is zero in linear problems. In simple cases where L is the difference between kinetic and potential energy, this proves the well-known result that the average kinetic and potential energies are equal.

It should be noted that the dispersion relation between ω and $\boldsymbol{\kappa}$ is not used in the course of the calculation of $\mathscr{L}$. The result in (12) shows clearly that this is crucial. If the dispersion relation were used, the class of functions would be too restrictive and we would obtain merely the final stationary value $\mathscr{L} = 0$. In nonlinear problems, this separation of the dispersion relation from the *form* of the periodic solution is more complicated. It is discussed in general in § 7.

The second special case is a simple example to illustrate these nonlinear effects. It is also a useful model on which to focus the general discussion of the perturbation methods. It is a nonlinear version of the Klein–Gordon equation, namely, $u_{tt} - u_{xx} + V'(u) = 0$. The Lagrangian is

$$L = \tfrac{1}{2}u_t^2 - \tfrac{1}{2}u_x^2 - V(u). \tag{14}$$

The periodic solution $u = U_0(\theta)$ satisfies

$$(\omega^2 - \kappa^2)U_0''(\theta) + V'(U_0) = 0, \tag{15}$$

which has the integral

$$\tfrac{1}{2}(\omega^2 - \kappa^2)U_0'^2 + V(U_0) = A. \tag{16}$$

The constant of integration A can be used as a parameter equivalent to the amplitude a. Equation (16) can be solved to give the function $U_0(\theta)$ in the inverse form

$$\theta = [\tfrac{1}{2}(\omega^2 - \kappa^2)]^{1/2} \int \frac{dU_0}{[A - V(U_0)]^{1/2}}. \tag{17}$$

Since θ is normalized so that the period is 2π, it follows that

$$2\pi = [\tfrac{1}{2}(\omega^2 - \kappa^2)]^{1/2} \oint \frac{dU_0}{[A - V(U_0)]^{1/2}}, \tag{18}$$

where $\oint$ denotes integration over a complete period. This is the dispersion relation between ω, κ and A.

The averaged Lagrangian is first introduced directly from (14) as

$$\bar{L} = \frac{1}{2\pi} \int_0^{2\pi} \{\tfrac{1}{2}(\omega^2 - \kappa^2)U_0'^2 - V(U_0)\}\, d\theta. \tag{19}$$

We then use (16) to obtain a function of ω, κ, A in the following way.

We have successively

$$
\begin{aligned}
\bar{L} &= \frac{1}{2\pi}\int_0^{2\pi} (\omega^2 - \kappa^2)U_0'^2\,d\theta - A \\
&= \frac{1}{2\pi}\int_0^{2\pi} (\omega^2 - \kappa^2)U_0'\,dU_0 - A \qquad (20)\\
&= \frac{1}{2\pi}[2(\omega^2 - \kappa^2)]^{1/2}\oint [A - V(U_0)]^{1/2}\,dU_0 - A.
\end{aligned}
$$

In this final form the U_0 has become merely a dummy variable of integration and we have a definite function of ω, κ, A. The symbol $\mathscr{L}(\omega, \kappa, A)$ is now reserved for this final form. The function $\mathscr{L}(\omega, \kappa, A)$ is used in (9) and (10). It is observed immediately that the dispersion relation (18) is just $\mathscr{L}_A = 0$, in accordance with the general result.

The important point about the manipulation in (20) is that only the equation (16) is used. Although, as noted in (17) and (18), this equation contains the solution for U_0 and the dispersion relation, these are not used explicitly. If they were, it would be possible to restrict the averaged Lagrangian too much to use its variation. Then (9) and (10) would not follow. For example, the dispersion relation (18) must not be used to eliminate $\omega^2 - \kappa^2$ from (20) altogether. We shall see later, in § 7, how to describe the manipulation in general terms and how to justify the results.

3. Two-timing on the Euler equations. The so-called "two-timing" method recognizes explicitly in the expressions for the dependent variables that changes are occurring on two time scales: the "fast" oscillations of the wave train and the "slow" variations of the parameters $(\omega, \boldsymbol{\kappa}, a)$. There are two corresponding length scales. The method originated in the study of the ordinary differential equations governing the nonlinear vibrations of mechanical systems, where it was apparently first introduced by Krylov and Bogoliubov. It has received considerable development recently (see Cole [**1968**]). Luke [**1966**] adapted it to the partial differential equations of wave problems using a basic paper by Kuzmak [**1959**] as a source.

The main idea is to express $u(\boldsymbol{x}, t)$ in the form

$$
u(\boldsymbol{x}, t) = U(\theta, \boldsymbol{X}, T, \varepsilon), \qquad (21)
$$

where

$$
\theta = \varepsilon^{-1}\Theta(\boldsymbol{X}, T), \qquad \boldsymbol{X} = \varepsilon\boldsymbol{x}, \qquad T = \varepsilon t, \qquad (22)
$$

and the small parameter ε measures the ratio of the fast time scale to the slow time scale. If the wave-number $\boldsymbol{\kappa}$ and frequency ω are introduced, we have

$$\omega(X, T) = -\theta_t = -\Theta_T, \quad \boldsymbol{\kappa}(X, T) = \theta_{\boldsymbol{x}} = \Theta_{\boldsymbol{x}}. \tag{23}$$

The scaling has been arranged so that

$$\frac{\partial \omega}{\partial t} = \varepsilon \frac{\partial \omega}{\partial T}, \qquad \frac{\partial \omega}{\partial x_i} = \varepsilon \frac{\partial \omega}{\partial X_i}, \tag{24}$$

with similar expressions for $\boldsymbol{\kappa}$, and

$$\frac{\partial u}{\partial t} = -\omega \frac{\partial U}{\partial \theta} + \varepsilon \frac{\partial U}{\partial T}, \qquad \frac{\partial u}{\partial x_i} = \kappa_i \frac{\partial U}{\partial \theta} + \varepsilon \frac{\partial U}{\partial X_i}. \tag{25}$$

If $\boldsymbol{X}, T, \omega, \boldsymbol{\kappa}, U$ are all taken to be $O(1)$ quantities, the scaling has been arranged so that $\omega, \boldsymbol{\kappa}$ are slowly varying quantities, and so that u has a slow variation in addition to its oscillation with the phase θ.

The basis of the technique is to use the function $U(\theta, \boldsymbol{X}, T, \varepsilon)$ explicitly as a function of all the independent variables $\theta, \boldsymbol{X}, T$ even though ultimately the extra variable θ is to be related back to $\boldsymbol{X}$ and T by (22). This extra freedom can be used to advantage in obtaining a uniformly valid expansion for u.

The relevant differential equation in the present case is the Euler equation (2). For simplicity in the discussion, the case of one space dimension x is considered; the extension is trivial. The first step is to convert the equation (2) into an equation for[2] $U(\theta, X, T)$. In doing this, we must use $\theta = \varepsilon^{-1}\Theta(X, T)$. As in (25), t and x derivatives become

$$\frac{\partial}{\partial t} = \Theta_T \frac{\partial}{\partial \theta} + \varepsilon \frac{\partial}{\partial T}, \qquad \frac{\partial}{\partial x} = \Theta_X \frac{\partial}{\partial \theta} + \varepsilon \frac{\partial}{\partial X}. \tag{26}$$

To preserve the symmetry between x and t it will be convenient to work with $\nu = \Theta_T = -\omega$ rather than ω. Then, (2) becomes

$$\frac{\partial}{\partial \theta}(\nu L_1 + \kappa L_2) - L_3 + \varepsilon \frac{\partial L_1}{\partial T} + \varepsilon \frac{\partial L_2}{\partial X} = 0. \tag{27}$$

In L_1, L_2, L_3, which are defined in terms of L by (3), the arguments are as shown:

$$L_j = L_j(\nu U_\theta + \varepsilon U_T, \kappa U_\theta + \varepsilon U_X, U). \tag{28}$$

At this point, the relation of θ to X, T may be dropped temporarily, and

[2] The dependence of U on the small parameter ε will not be shown, whenever it is not a point to be emphasized.

(27) can be considered as an equation for the function $U(\theta, X, T)$ of the three *independent* variables θ, X, T. Clearly, if this function is found, then $U(\varepsilon^{-1}\Theta, X, T)$ solves the original problem. It should be noted that $\Theta(X, T)$ still appears in (27), through $\nu = \Theta_T, \kappa = \Theta_X$, but the relation of Θ to the argument θ in U is dropped.

The solution of (27) is now obtained by expanding U formally in a power series

$$U(\theta, X, T, \varepsilon) = \sum_{n=0}^{\infty} \varepsilon^n U_n(\theta, X, T),$$

and equating the terms of successive orders in ε to zero. To lowest order in ε, only U_0 is involved and we have

$$\frac{\partial}{\partial \theta}\{\nu L_1^{(0)} + \kappa L_2^{(0)}\} - L_3^{(0)} = 0, \tag{29}$$

where

$$L_j^{(0)} = L_j(\nu U_{0\theta}, \kappa U_{0\theta}, U_0). \tag{30}$$

Since (29), (30) involve only θ derivatives of U_0, this is effectively an ordinary differential equation, in fact that of the uniform periodic wave train. In solving it an arbitrary "constant" A will arise which is equivalent to the amplitude a. Of course, this "constant" must now be allowed to be a function of (X, T). Just the amplitude modulation required!

At this lowest order, U_0 will be determined in its dependence on θ, but will involve the three parameters ν, κ, A which are all functions of X, T. There is one relation between them at this stage. In solving for U_0 and normalizing the period in θ to 2π, the parameters ν, κ, A will be required to satisfy the dispersion relation; in this direct attack on the Euler equations there are no subtleties about whether the dispersion relation is or is not used. In the Klein–Gordon example (14), the lowest-order equation (29) is just (15). It integrates to (16), now with $A = A(X, T)$, the dependence of U_0 on θ is determined, together with the dispersion relation as in (17), (18).

Further relations between ν, κ, A are obtained by going to the next order in ε. The next-order terms in (27) involve U_1, U_0, ν, κ. Only θ derivatives of U_1 occur, so it is again, effectively, an ordinary differential equation for U_1. The solution will not be periodic in θ; there will be "secular" terms linear in θ, unless an appropriate condition is enforced on the nonhomogeneous terms. For the expansion of U to be uniformly valid in θ, the secular terms *must* be suppressed. We shall call this requirement a "secular condition". It is similar to the orthogonality condition required of the nonhomogeneous terms in certain linear problems of eigenfunction expansions. The secular condition gives a further equation for ν, κ, A.

The suppression of secular terms is most easily carried out on conservation equations. Indeed the whole two-timing procedure is considerably simplified by working with conservation equations, and some of the complication Luke ran into is avoided.

The first step is to write (27) in the equivalent conservation form[3]

$$\frac{\partial}{\partial \theta}\{(\nu L_1 + \kappa L_2)U_\theta - L\} + \varepsilon\frac{\partial}{\partial T}(U_\theta L_1) + \varepsilon\frac{\partial}{\partial X}(U_\theta L_2) = 0. \tag{31}$$

We shall denote this by

$$\partial R/\partial \theta + \varepsilon \partial P/\partial T + \varepsilon \partial Q/\partial X = 0 \tag{32}$$

for ease of discussion. The quantities P, Q, R are functions of U, U_θ, U_X, U_T, ν, κ and ε. If we now expand U in a power series in ε, the quantities P, Q, R will have corresponding expansions

$$P = \sum_0^\infty P^{(n)}\varepsilon^n, \quad \text{etc.}$$

The lowest terms $P^{(0)}$, $Q^{(0)}$, $R^{(0)}$ depend on U_0, ν, κ; the next terms $P^{(1)}, Q^{(1)}, R^{(1)}$ includes also U_1.

The first two terms of (32) are

$$\frac{\partial R^{(0)}}{\partial \theta} = 0, \tag{33}$$

$$\frac{\partial R^{(1)}}{\partial \theta} = -\frac{\partial P^{(0)}}{\partial T} - \frac{\partial Q^{(0)}}{\partial X}. \tag{34}$$

The first one is equivalent to (29) and leads to the immediate first integral

$$R^{(0)} \equiv (\nu L_1^{(0)} + \kappa L_2^{(0)})U_{0\theta} - L^{(0)} = A(X, T). \tag{35}$$

(This is (16) for the Klein–Gordon example.) The second equation (34) has to be solved for U_1, which occurs in $R^{(1)}$. A solution uniformly valid in θ requires U, and hence each U_n, to be *periodic* in θ with period 2π. This is only possible in (34) if the integral of the right-hand side with respect to θ over one period is zero. That is, to avoid secular terms, we must demand

$$\frac{\partial}{\partial T}\left(\frac{1}{2\pi}\int_0^{2\pi} P^{(0)}\, d\theta\right) + \frac{\partial}{\partial X}\left(\frac{1}{2\pi}\int_0^{2\pi} Q^{(0)}\, d\theta\right) = 0. \tag{36}$$

From the definitions of P, Q, R (see (31) and (32)), it is

$$\frac{\partial}{\partial T}\left(\frac{1}{2\pi}\int_0^{2\pi} U_{0\theta}L_1^{(0)}\, d\theta\right) + \frac{\partial}{\partial X}\left(\frac{1}{2\pi}\int_0^{2\pi} U_{0\theta}L_2^{(0)}\, d\theta\right) = 0. \tag{37}$$

[3] A natural derivation of this equation, free of ingenuity, will be noted later in § 5.

This is the remaining equation for ν, κ, A.

In this approach, (35) is solved completely for U_0 and the dispersion relation. The results are substituted in (37). The tie-in with the averaged Lagrangian raises the more subtle question of just how (35) is to be used in conjunction with (37). We may note, however, that the averaged Lagrangian is

$$\overline{L^{(0)}} = \frac{1}{2\pi}\int_0^{2\pi} L(\nu U_{0\theta}, \kappa U_{0\theta}, U_0)\, d\theta. \tag{38}$$

If derivatives are taken, keeping U_0 fixed, then (37) can indeed be written

$$\frac{\partial}{\partial T}\overline{L_\nu^{(0)}} + \frac{\partial}{\partial X}\overline{L_\kappa^{(0)}} = 0. \tag{39}$$

The form $\mathscr{L}(\nu, \kappa, A)$, with appropriate careful use of (35) in (38), still gives

$$\frac{\partial}{\partial T}\mathscr{L}_\nu + \frac{\partial}{\partial X}\mathscr{L}_\kappa = 0, \tag{40}$$

even though further dependence on ν and κ has been introduced into (38) via U_0. Moreover, the dispersion relation becomes $\mathscr{L}_A = 0$. This preferred form is derived in § 7 after the averaged variational principle has been justified and used directly to derive the results of this section.

For the Klein–Gordon example, L is given by (14), so that

$$L^{(0)} = \tfrac{1}{2}(\nu^2 - \kappa^2)U_{0\theta}^2 - V(U_0)$$

and (39) becomes

$$\frac{\partial}{\partial T}\left(\frac{\nu}{2\pi}\int_0^{2\pi} U_{0\theta}^2\, d\theta\right) - \frac{\partial}{\partial X}\left(\frac{\kappa}{2\pi}\int_0^{2\pi} U_{0\theta}^2\, d\theta\right) = 0. \tag{41}$$

On the other hand, from (20), $\mathscr{L}$ is

$$\mathscr{L} = \frac{1}{2\pi}[2(\nu^2 - \kappa^2)]^{1/2}\oint [A - V(U_0)]^{1/2}\, dU_0 - A, \tag{42}$$

and (40) becomes

$$\begin{aligned}&\frac{\partial}{\partial T}\left(\frac{1}{2\pi}\frac{\nu}{[\nu^2 - \kappa^2]^{1/2}}\oint [A - V(U_0)]^{1/2}\, dU_0\right) \\ &\qquad - \frac{\partial}{\partial X}\left\{\frac{1}{2\pi}\frac{\kappa}{[\nu^2 - \kappa^2]^{1/2}}\oint [A - V(U_0)]^{1/2}\, dU_0\right\} = 0.\end{aligned} \tag{43}$$

The equivalence of these two forms of the secular condition, (41) and (43), is easily established for this example from (16).

Two-timing on the equations, then, is a satisfactory and consistent way to solve the problem. The expressions are kept relatively simple by writing the basic equations (27) and (31) in terms of the Lagrangian L. However, the full power of the variational principle has not been used in the two-timing analysis and the final results have not yet been justified in general in their most compact form using $\mathscr{L}$.

It is interesting to observe that an exact form of the secular condition can be derived from (32), valid without approximation for small ε, and hence valid for all orders. The only argument needed is that R is periodic in θ; therefore, from (32),

$$\frac{\partial}{\partial T}\frac{1}{2\pi}\int_0^{2\pi} P\,d\theta + \frac{\partial}{\partial X}\frac{1}{2\pi}\int_0^{2\pi} Q\,d\theta = 0. \tag{44}$$

Substituting for P and Q from (31), we have

$$\begin{aligned}&\frac{\partial}{\partial T}\frac{1}{2\pi}\int_0^{2\pi} U_\theta L_1(\nu U_\theta + \varepsilon U_T, \kappa U_\theta + \varepsilon U_X, U)\,d\theta \\ &\qquad + \frac{\partial}{\partial X}\frac{1}{2\pi}\int_0^{2\pi} U_\theta L_2(\nu U_\theta + \varepsilon U_T, \kappa U_\theta + \varepsilon U_X, U)\,d\theta = 0.\end{aligned} \tag{45}$$

The successive terms in the expansion of this for $U = \sum \varepsilon^n U_n$ give the required secular terms in the successive determinations of the U_n. Equation (37) is the lowest-order term in the expansion of (45).

4. Averaging and two-timing a system of conservation equations. In a first attack on slowly varying nonlinear waves (Whitham [**1965a**]), the equations for (ν, κ, A), and similar overall parameters that arise for higher-order systems, were obtained by averaging an equivalent system of conservation equations. The equivalent system is denoted by

$$\frac{\partial P_i}{\partial t} + \frac{\partial Q_i}{\partial x} = 0 \qquad (i = 1, \ldots, m). \tag{46}$$

The intuitive argument is to use their averaged form

$$\frac{\partial}{\partial t}\left(\frac{1}{2\pi}\int_0^{2\pi} P_i\,d\theta\right) + \frac{\partial}{\partial x}\left(\frac{1}{2\pi}\int_0^{2\pi} Q_i\,d\theta\right) = 0, \tag{47}$$

where the averages are computed from the periodic solution.

From the two-timing formalism, (46) becomes

$$\begin{aligned}&\left(\nu\frac{\partial}{\partial\theta} + \varepsilon\frac{\partial}{\partial T}\right)P_i + \left(\kappa\frac{\partial}{\partial\theta} + \varepsilon\frac{\partial}{\partial X}\right)Q_i = 0,\\ &\text{i.e.}\quad \frac{\partial R_i}{\partial\theta} + \varepsilon\frac{\partial P_i}{\partial T} + \varepsilon\frac{\partial Q_i}{\partial X} = 0, \qquad R_i = \nu P_i + \kappa Q_i.\end{aligned} \tag{48}$$

For each of these we then have

$$R_i^{(0)} = \nu P_i^{(0)} + \kappa Q_i^{(0)} = A_i(X, T), \tag{49}$$

and

$$\frac{\partial}{\partial T} \frac{1}{2\pi} \int_0^{2\pi} P_i^{(0)}\, d\theta + \frac{\partial}{\partial X} \frac{1}{2\pi} \int_0^{2\pi} Q_i^{(0)}\, d\theta = 0. \tag{50}$$

These correspond to (35) and (36), respectively. The integrals in (49) determine the periodic solution with the parameters ν, κ, A_i extended to allow dependence on (X, T). The secular conditions (50) complete the system of equations for ν, κ, A_i. This provides the formal justification for the more intuitive averaging in (47). Again it is observed from (48) that

$$\frac{\partial}{\partial T} \frac{1}{2\pi} \int_0^{2\pi} P_i\, d\theta + \frac{\partial}{\partial X} \frac{1}{2\pi} \int_0^{2\pi} Q_i\, d\theta = 0$$

is an exact secular condition.

The method was superseded by the more concise variational approach. This method based on conservation equations is only general if one can be sure that the requisite number of conservation equations (46) always exist. On specific problems, it was always found to be the case, and the experience was sufficiently varied to form the belief that this would always be the case. A direct proof can probably be found, but it is not an easy question. The next section which obtains the equations for ν, κ, A_i without this step implies an indirect verification.

5. Two-timing the variational principle. Most of the discussion will be given for a single function u as in (1), and for one space dimension x. The extensions will then be straightforward.

When u is written in the form (21), (22), the Euler equation becomes

$$\frac{\partial}{\partial \theta}(\nu L_1 + \kappa L_2) - L_3 + \varepsilon \frac{\partial L_1}{\partial T} + \varepsilon \frac{\partial L_2}{\partial X} = 0, \tag{51}$$

with

$$L_j = L_j(\nu U_\theta + \varepsilon U_T, \kappa U_\theta + \varepsilon U_X, U), \tag{52}$$

as noted in (27) and (28). It is surprising that this is just the Euler equation of the *three*-variable variational principle

$$\delta \iiint_0^{2\pi} L(\nu U_\theta + \varepsilon U_T, \kappa U_\theta + \varepsilon U_X, U)\, dT\, dX\, d\theta = 0, \tag{53}$$

for the *three*-variable function $U(\theta, X, T)$. In (53) the function U and its variations are taken to be periodic in θ, and the variations in U vanish on the boundary of the (X, T) region. It is even more surprising that (53) is already the averaged variational principle and it is *exact*! What started out as an intuitive argument for the first term of an approximate expansion is not only justified but turns out to contain the whole expansion. There is in fact no assumption that ε is small; the only step was to express $u(x, t)$ in the form (21), (22).

The exact averaged Lagrangian is defined by

$$\bar{L} = \frac{1}{2\pi}\int_0^{2\pi} L(\nu U_\theta + \varepsilon U_T, \kappa U_\theta + \varepsilon U_X, U)\, d\theta, \tag{54}$$

and (53) may be written

$$\delta \int\int \bar{L}\, dT\, dX = 0. \tag{55}$$

Variation of U in (53) gives the whole equation (51). But the simplicity of the arguments in § 2 depends on obtaining the equations for the slowly varying parameters by varying those parameters in the average variational principle. Accordingly, we consider variations of (53) with respect to the function $\Theta(X, T)$, which appears in (53) through $\nu = \Theta_T$, $\kappa = \Theta_X$. We have, first,

$$\int\int \{\bar{L}_\nu \delta\Theta_T + \bar{L}_\kappa \delta\Theta_X\}\, dX\, dT = 0,$$

using (54) for the inner integrals with respect to θ. By the usual variational argument we deduce

$$\frac{\partial}{\partial T}\bar{L}_\nu + \frac{\partial}{\partial X}\bar{L}_\kappa = 0, \tag{56}$$

for arbitrary $\delta\Theta$ which vanish on the boundary of the (X, T) region. This is just the secular condition (45), and it is also exact!

We are now free to use (53) for independent variations of $U(\theta, X, T)$ and $\Theta(X, T)$ to obtain both the equations for U and the secular conditions.

To lowest order, we have

$$\delta \int\int\int_0^{2\pi} L(\nu U_{0\theta}, \kappa U_{0\theta}, U_0)\, dT\, dX\, d\theta = 0. \tag{57}$$

The variation with respect to U_0 gives

$$\frac{\partial}{\partial\theta}(\nu L_1^{(0)} + \kappa L_2^{(0)}) - L_3^{(0)} = 0 \tag{58}$$

(using the notation of (30)) and the variation with respect to Θ gives

$$\frac{\partial}{\partial T}\overline{L_{\nu}^{(0)}} + \frac{\partial}{\partial X}\overline{L_{\kappa}^{(0)}} = 0. \tag{59}$$

These are the results (29) and (39) of § 3 obtained very simply. Essentially this justifies the theory of § 2; it remains to discuss the various forms of the averaged Lagrangian. This is completed in § 7, after further amplification of the theory thus far.

The conservation equation (31), and its lowest approximation used in obtaining the integral (35), follow naturally from (53) and (57). In each case the variable θ does not appear explicitly in L. Therefore, the variational principle is invariant if an arbitrary constant is added to θ. Noether's theorem proves that there is a conservation equation corresponding to this invariance; it is

$$\frac{\partial}{\partial\theta}(U_\theta L_{U_\theta} - L) + \frac{\partial}{\partial T}(U_\theta L_{U_T}) + \frac{\partial}{\partial X}(U_\theta L_{U_X}) = 0. \tag{60}$$

With the arguments of L given in (53) it is easily seen that (31) follows. The arguments in (57) give the lowest-order approximation (35). We no longer need (31) for the derivation of the secular condition, but (35) is useful in connection with (58).

Extension to more variables. The result extends to a system with dependent variables $u_i(\boldsymbol{x}, t)\ (i = 1, \ldots, m)$, and with more space dimensions. If the system can be derived from the variational principle

$$\delta \int\int L(u_{it}, u_{i\boldsymbol{x}}, u_i)\, d\boldsymbol{x}\, dt = 0, \tag{61}$$

the Euler equations are

$$\frac{\partial}{\partial t}L_{i1} + \frac{\partial}{\partial \boldsymbol{x}}L_{i2} - L_{i3} = 0 \qquad (i = 1, \ldots, m)$$

where

$$L_{i1} = \partial L/\partial u_{it}, \qquad L_{i2} = \partial L/\partial u_{i\boldsymbol{x}}, \qquad L_{i3} = \partial L/\partial u_i.$$

After two-timing all the u's in the form

$$u_i(\boldsymbol{x}, t) = U_i(\theta, \boldsymbol{X}, T), \qquad \theta = \varepsilon^{-1}\Theta(\boldsymbol{X}, T),$$

the Euler equations become

$$\frac{\partial}{\partial\theta}(\nu L_{i1} + \kappa L_{i2}) - L_{i3} + \varepsilon\frac{\partial}{\partial T}L_{i1} + \varepsilon\frac{\partial}{\partial \boldsymbol{X}}L_{i2} = 0,$$

where the arguments of the L_{i1}, L_{i2}, L_{i3} are

$$u_{it} = \nu U_{i\theta} + \varepsilon U_{iT}, \qquad u_{ix} = \kappa U_{i\theta} + \varepsilon U_{i\mathbf{X}}, \qquad u_i = U_i.$$

These are the Euler equations for

$$\delta \int\int\int_0^{2\pi} L(\nu U_{i\theta} + \varepsilon U_{iT}, \boldsymbol{\kappa} U_{i\theta} + \varepsilon U_{i\mathbf{X}}, U_i)\, d\mathbf{X}\, dT\, d\theta = 0. \tag{62}$$

In applications of these methods to such higher-order systems, it was noted in earlier papers that some of the u_i appear only through their derivatives in (61). Typically, such u_i are potentials, whose derivatives are physical quantities. Let one of these be denoted by ϕ. In the uniform periodic wave train, for maximum generality ϕ must be expressed as

$$\phi = \beta x + \gamma t + \Phi(\theta), \qquad \theta = \kappa x + \nu t, \tag{63}$$

where $\Phi(\theta)$ is periodic and β, γ are arbitrary constants. The derivatives of ϕ, which are the physical quantities, are then periodic functions of θ. The pair (β, γ) behaves like (κ, ν) in the analysis. They have been called pseudo wavenumbers and frequencies, since they give the increments in ϕ over one period or wavelength, just as (κ, ν) give the increments of θ.

In the slowly varying wave train, (63) is generalized to

$$\begin{aligned} &\phi = \psi + \Phi(\theta, X, T), \\ &\theta = \varepsilon^{-1}\Theta(X, T), \qquad \psi = \varepsilon^{-1}\Psi(X, T), \\ &\nu = \Theta_T, \qquad \kappa = \Theta_X, \qquad \gamma = \Psi_T, \qquad \beta = \Psi_X. \end{aligned} \tag{64}$$

The average Lagrangian in (62) now includes dependence on $\Psi(X, T)$ through $\gamma = \Psi_T, \beta = \Psi_X$. The variations with respect to Ψ give additional secular conditions

$$\frac{\partial}{\partial T}\bar{L}_\gamma + \frac{\partial}{\partial X}\bar{L}_\beta = 0, \tag{65}$$

and these are required to complete the solution to lowest order. The details are not given here in general; typical cases can be seen in the previous papers (Whitham [**1965**], [**1967**]) and will be included in the examples of § 8.

6. Relation between two-timing and averaging. It is intriguing to investigate further the relation of the precise result (53) with the intuitive ideas of averaging. One might also question whether (53) can be posed directly without any reference to the Euler equations, even though their use in § 5 has been held to a minimum.

With the choice of the functional form (21) and (22) the original variational principle (1) becomes

$$(66) \qquad \delta \int\int [L(\nu U_\theta + \varepsilon U_T, \kappa U_\theta + \varepsilon U_X, U)]_{\theta = \varepsilon^{-1}\Theta} \, dX \, dT = 0,$$

where $[\ ]_{\theta=\varepsilon^{-1}\Theta}$ denotes that, at this first step, θ is still linked to (X, T) through the substitution $\theta = \varepsilon^{-1}\Theta(X, T)$. The idea of the averaging process is that the dependence on ε through $\theta = \varepsilon^{-1}\Theta$ is highly oscillatory so that the integrand in (66) may be replaced by its mean value. A consistent way to introduce this is to use Fourier series. For functions $U(\theta, X, T, \varepsilon)$ periodic in θ, $L(\nu U_\theta + \varepsilon U_T, \kappa U_\theta + \varepsilon U_X, U)$ is also periodic in θ and we may introduce its Fourier series

$$(67) \quad L(\nu U_\theta + \varepsilon U_T, \kappa U_\theta + \varepsilon U_X, U) = C_0 + \sum_{n=1}^{\infty} (C_n \cos n\theta + S_n \sin n\theta).$$

The coefficients C_n, S_n are functionals of $U(\theta, X, T, \varepsilon)$, and of $\Theta(X, T)$ through $\nu = \Theta_T$, $\kappa = \Theta_X$. In detail

$$(68) \qquad C_0 = \frac{1}{2\pi} \int_0^{2\pi} L(\nu U_\theta + \varepsilon U_T, \kappa U_\theta + \varepsilon U_X, U) \, d\theta,$$

and

$$(69) \qquad C_n = \frac{1}{\pi} \int_0^{2\pi} L \cos n\theta \, d\theta, \qquad S_n = \frac{1}{\pi} \int_0^{2\pi} L \sin n\theta \, d\theta,$$

(with the same arguments for L). Then, (66) becomes

$$(70) \qquad \delta \int\int \left\{ C_0 + \sum_{n=1}^{\infty} \left(C_n \cos \frac{n\Theta}{\varepsilon} + S_n \sin \frac{n\Theta}{\varepsilon} \right) \right\} dX \, dT = 0.$$

Comparing (53) and (68), we see that (53) is in fact

$$(71) \qquad \delta \int\int C_0 \, dX \, dT = 0.$$

Since (71) is therefore exact, it must be valid to omit the other Fourier terms in (70).

The basis of the intuitive argument is that, for small ε, the oscillatory terms in the integral in (70) contribute very little and may be omitted. Indeed, by repeated integration by parts they may be made of smaller order than *any* power of ε. If they are then neglected in (70) before taking the variations, we have (71) of course. However, if they are retained, they now involve high derivatives in Θ and U, and the usual variational argument to establish the Euler equations will undo the integration by parts and we are back to (70). Moreover, variations of Θ in (70) will include terms of order ε^{-1}, from the variations of $\cos n\varepsilon^{-1}\Theta$ and $\sin n\varepsilon^{-1}\Theta$, and these would appear to be the dominant terms. Yet we know already that (71) is correct.

There is then an apparent lack of uniqueness with respect to the ordering of the terms in (70) with respect to ε. The only way out of this disturbing situation would be if each of the Fourier terms in (70) is individually stationary, i.e.

$$\delta \int\int \left(C_n \cos \frac{n\Theta}{\varepsilon} + S_n \sin \frac{n\Theta}{\varepsilon} \right) dX\, dT = 0, \tag{72}$$

for all n. Surprisingly enough this turns out to be true.

To prove it, we show that the Euler equations for (72), as well as those for (71), are equivalent to (51). The coefficients C_n, S_n are defined by (69). Accordingly (72) may be written

$$\delta \int\int\int_0^{2\pi} \cos n\left(\theta - \frac{\Theta}{\varepsilon}\right) L(\nu U_\theta + \varepsilon U_T, \kappa U_\theta + \varepsilon U_X, U)\, dX\, dT\, d\theta = 0. \tag{73}$$

The Euler equation for variations in U is

$$\begin{aligned} &\frac{\partial}{\partial \theta}\left\{(\nu L_1 + \kappa L_2) \cos n\left(\theta - \frac{\Theta}{\varepsilon}\right)\right\} - L_3 \cos n\left(\theta - \frac{\Theta}{\varepsilon}\right) \\ &\quad + \varepsilon \frac{\partial}{\partial T}\left\{L_1 \cos n\left(\theta - \frac{\Theta}{\varepsilon}\right)\right\} + \varepsilon \frac{\partial}{\partial X}\left\{L_2 \cos n\left(\theta - \frac{\Theta}{\varepsilon}\right)\right\} = 0. \end{aligned} \tag{74}$$

This equation may be expanded to

$$\left\{\frac{\partial}{\partial \theta}(\nu L_1 + \kappa L_2) - L_3 + \varepsilon \frac{\partial L_1}{\partial T} + \varepsilon \frac{\partial L_2}{\partial X}\right\} \cos n\left(\theta - \frac{\Theta}{\varepsilon}\right) = 0. \tag{75}$$

The expression in brackets is just (51) and the result is proved.

Again all the information is in the equation for U, but we may consider the variations of (73) with respect to Θ. The result is

$$\begin{aligned} &\frac{\partial}{\partial T} \frac{1}{2\pi} \int_0^{2\pi} U_\theta L_1 \cos n\left(\theta - \frac{\Theta}{\varepsilon}\right) d\theta + \frac{\partial}{\partial X} \int_0^{2\pi} U_\theta L_2 \cos n\left(\theta - \frac{\Theta}{\varepsilon}\right) d\theta \\ &\qquad - \frac{1}{2\pi} \int_0^{2\pi} \frac{nL}{\varepsilon} \sin n\left(\theta - \frac{\Theta}{\varepsilon}\right) d\theta = 0. \end{aligned}$$

It may be expanded to read

$$\left\{\varepsilon \frac{\partial P_n^c}{\partial T} + \varepsilon \frac{\partial Q_n^c}{\partial X} + nR_n^s\right\} \cos \frac{n\Theta}{\varepsilon} + \left\{\varepsilon \frac{\partial P_n^s}{\partial T} + \varepsilon \frac{\partial Q_n^s}{\partial X} - nR_n^c\right\} \sin \frac{n\Theta}{\varepsilon} = 0, \tag{76}$$

where P, Q, R refer to the quantities introduced in (31), (32) and superscripts P_n^c, P_n^s, etc., denote the cosine and sine coefficients in their Fourier series. It is easily shown from (32), by a continuation of the secular argument, that the bracketed terms in (76) vanish separately. For the secular condition is that P, Q, R must be periodic in θ. Hence, these quantities can be expanded in their Fourier series

$$P = P_0 + \sum_{n=1}^{\infty} (P_n^c \cos n\theta + P_n^s \sin n\theta), \quad \text{etc.} \tag{77}$$

Substitution of these in (32) gives

$$\frac{\partial P_0}{\partial T} + \frac{\partial Q_0}{\partial X} = 0, \qquad \varepsilon\frac{\partial P_n^c}{\partial T} + \varepsilon\frac{\partial Q_n^c}{\partial X} = -nR_n^s, \qquad \varepsilon\frac{\partial P_n^s}{\partial T} + \varepsilon\frac{\partial Q_n^s}{\partial X} = nR_n^c. \tag{78}$$

The first of these is the main secular condition noted in (44); the additional ones add nothing new in the lowest order.

7. Optimum form of the average Lagrangian. The variational principle (57) and its consequences in (58) and (59) have been justified. We now consider the various manipulations in order to obtain the most effective form. The discussion throughout this section will refer to the lowest-order approximation U_0 and the subscript zero will be dropped; the corresponding superscript zero will be dropped in (58) and (59).

In the first approximation, then, we have

$$\delta \int\int \bar{L}\, dX\, dT = 0, \tag{79}$$

where

$$\bar{L} = \frac{1}{2\pi}\int_0^{2\pi} L(\nu U_\theta, \kappa U_\theta, U)\, d\theta, \qquad \nu = \Theta_T, \kappa = \Theta_X. \tag{80}$$

The variation with respect to $U(\theta, X, T)$ gives

$$\frac{\partial}{\partial\theta}(\nu L_1 + \kappa L_2) - L_3 = 0; \tag{81}$$

the variation with respect to $\Theta(X, T)$ gives

$$\frac{\partial}{\partial T}\bar{L}_\nu + \frac{\partial}{\partial X}\bar{L}_\kappa = 0. \tag{82}$$

Equation (81) has the integral

$$(\nu L_1 + \kappa L_2)U_\theta - L = A(X, T). \tag{83}$$

The aim is to use this integral to evaluate $\bar{L}$ as a function of (ν, κ, A).

The method is essentially a Hamiltonian version of the equations. The quantity U_θ is eliminated in favour of $\partial L/\partial U_\theta$ just as $\dot{q}$ is eliminated in favour of a generalized momentum $p = \partial L/\partial \dot{q}$ in ordinary dynamics. A new variable Π is defined by

$$\Pi = \partial L/\partial U_\theta = \nu L_1 + \kappa L_2, \tag{84}$$

and the "Hamiltonian" $H(\Pi, U; \Theta)$ is defined by

$$H = U_\theta \partial L/\partial U_\theta - L = U_\theta(\nu L_1 + \kappa L_2) - L. \tag{85}$$

It is noticed immediately that (83) is the "energy integral"

$$H(\Pi, U; \Theta) = A(X, T). \tag{86}$$

The Hamiltonian is a function of Π and U, but it is a functional of Θ since it depends on $\nu = \Theta_T$ and $\kappa = \Theta_X$. The semicolon notation will be used in displaying the arguments to make this distinction.

From the transformation defined by (84) and (85),

$$U_\theta = \partial H/\partial \Pi, \tag{87}$$

and (81) becomes

$$\Pi_\theta = -\partial H/\partial U. \tag{88}$$

These are the "Hamiltonian equations" to replace "Lagrange's equation" (81). The variational principle (79) may now be written with

$$\bar{L} = \frac{1}{2\pi}\int_0^{2\pi} (\Pi U_\theta - H)\, d\theta. \tag{89}$$

The *independent* variations of Π and U lead to the equations (87) and (88), respectively. The independence of the variations of Π and U is crucially important. It is an extension of the original form, because in (79), (80) the variation δU_θ, and hence $\delta\Pi$, is determined in terms of δU. Yet, we see that (87) and (88) do follow from (89), with $\delta\Pi$ and δU independent. Thus, we may use the extended form. It is exactly the extension used in ordinary mechanics.

We also saw that the stationary values of Π and U satisfy the integral (86). We may, therefore, take the stationary value for (89) from the class of functions Π, U that satisfy (86). It is no longer too restrictive to use (86) in its entirety, because the dispersion relation cannot be inferred without also using the relation of Π to U_θ. The latter is given by (87) and we do not use that. It should be emphasized again that this is only possible by the extension that Π and U can be varied independently in (89).

Finally, then, we solve (86) in the form

$$\Pi = \Pi(U, A; \Theta)$$

and (89) may be written

$$\bar{L} = \mathscr{L}\{A; \Theta\} = \frac{1}{2\pi}\oint \Pi \, dU - A. \tag{90}$$

$\mathscr{L}\{A; \Theta\}$ is a functional of Θ and a function of A; it is a function $\mathscr{L}(\nu, \kappa, A)$ of $\nu = \Theta_T$, $\kappa = \Theta_X$ and A.

Extension to higher-order systems. The generalization to cases with more functions u_i is carried out similarly. The exact averaged Lagrangian is given in (62). As noted in (64) some of the u_i may be potentials and require that two-timing form. It will suffice to explain the situation for higher-order systems in the case of two functions u_i, one of which is a potential. They will be denoted by u and ϕ. The Lagrangian is

$$L = L(u_t, u_x, u, \phi_t, \phi_x)$$

with no explicit dependence on ϕ. To lowest order the variational principle is (79) with

$$\bar{L} = \frac{1}{2\pi}\int_0^{2\pi} L(\nu U_\theta, \kappa U_\theta, U, \gamma + \nu\Phi_\theta, \beta + \kappa Q_\theta)\, d\theta, \tag{91}$$

$$\nu = \Theta_T, \qquad \kappa = \Theta_X, \qquad \gamma = \Psi_T, \qquad \beta = \Psi_X. \tag{92}$$

In addition to (81), we have the Euler equation from variations in Φ:

$$\frac{\partial}{\partial\theta}(\nu L_4 + \kappa L_5) = 0. \tag{93}$$

The absence of explicit dependence on Φ in (91) is significant in that (93) integrates immediately, and provides a second integral to use with (83). The variations of Ψ give a second secular condition

$$\frac{\partial}{\partial T}\bar{L}_\gamma + \frac{\partial}{\partial X}\bar{L}_\beta = 0 \tag{94}$$

to add to (82).

The Hamiltonian form is obtained from the transformation

$$\Pi_1 = \frac{\partial L}{\partial U_\theta} = \nu L_1 + \kappa L_2, \tag{95}$$

$$\Pi_2 = \frac{\partial L}{\partial \Phi_\theta} = \nu L_4 + \kappa L_5, \tag{96}$$

$$H(\Pi_1, \Pi_2, U, \Phi; \Theta, \Psi) = \Pi_1 U_\theta + \Pi_2 \Phi_\theta - L. \tag{97}$$

The Euler equations become

$$(98)\quad U_\theta = \frac{\partial H}{\partial \Pi_1}, \qquad \Phi_\theta = \frac{\partial H}{\partial \Pi_2}, \qquad \Pi_{1\theta} = -\frac{\partial H}{\partial U}, \qquad \Pi_{2\theta} = -\frac{\partial H}{\partial \Phi},$$

and follow from (91) with

$$(99)\qquad \bar{L} = \frac{1}{2\pi}\int_0^{2\pi} (\Pi_1 U_\theta + \Pi_2 \Phi_\theta - H)\, d\theta.$$

Again a crucial extension is that Π_1, Π_2, U, Φ may be varied independently.

Since Φ was in fact absent from L, it is also absent from H and the last equation in (98) gives

$$(100)\qquad \Pi_2 = B(X, T).$$

This is just the integral obtained from (93). The other one is $H = A(X, T)$, which (with (100)) can be solved to give

$$(101)\qquad \Pi_1 = \Pi_1(U, A, B; \Theta, \Psi).$$

In view of (100), the term

$$(102)\qquad \int_0^{2\pi} \Pi_2 \Phi_\theta\, d\theta = B \int_0^{2\pi} \Phi_\theta\, d\theta = 0$$

in (99), since Φ is periodic. Hence,

$$(103)\qquad \bar{L} = \mathscr{L}\{A, B; \Theta, \Psi\} = \frac{1}{2\pi}\oint \Pi_1\, dU - A,$$

where Π_1 is given by (101). The derivatives of Θ, Ψ appear in $\mathscr{L}$.

The required equations are then

$$(104)\qquad \mathscr{L}_A = 0, \qquad \mathscr{L}_B = 0,$$

$$(105)\qquad \frac{\partial}{\partial T}\mathscr{L}_\nu + \frac{\partial}{\partial X}\mathscr{L}_\kappa = 0, \qquad \frac{\partial}{\partial T}\mathscr{L}_\gamma + \frac{\partial}{\partial X}\mathscr{L}_\beta = 0.$$

Since the additional variable ϕ was taken to be a potential, it automatically brought in a second integral to simplify (99). In more general systems, one might expect sums of terms like $\Pi_1 U_\theta$ and $\Pi_2 \Phi_\theta$ in (99). All the Π_2 will be integrals and the terms $\Pi_2 \Phi_\theta$ integrate to zero as in (102). We then have

$$(106)\qquad \bar{L} = \frac{1}{2\pi}\int_0^{2\pi} \{\Sigma \Pi_1 U_\theta\}\, d\theta - A, \qquad \Pi_2 = B, \qquad H = A.$$

Further simplification down to integrals of the type in (90) would depend on finding further integrals. In all known examples, all the u_i except one are potentials. So there is only one U and Π_1; the form in (103) applies. Even if eventually other types arise, the final simplification is not essential; the method still applies.

Linear problems. For linear problems, one need not resort to this transformation to Hamiltonian form. Its object is to introduce some restriction on the *form* of the periodic solution into $\bar{L}$ while leaving enough flexibility to apply the variational arguments. In linear problems, the U_i are sinusoidal in θ. One can introduce the appropriate forms for them without being overly restrictive, provided the dispersion relation is not used. In nonlinear problems, the form of the U, the dispersion relation and the amplitudes are coupled together. These require the more ingenious treatment of this section.

8. Illustrative examples.

(a) *The Klein–Gordon equation.* The Lagrangian is

$$L = \tfrac{1}{2}u_t^2 - \tfrac{1}{2}u_x^2 - V(u),$$

and the lowest-order approximation is

$$L = \tfrac{1}{2}(\nu^2 - \kappa^2)U_\theta^2 - V(U).$$

The transformation (84), (85) to Hamiltonian form is

$$\begin{aligned} \Pi &= \partial L/\partial U_\theta = (\nu^2 - \kappa^2)U_\theta, \\ H &= U_\theta \partial L/\partial U_\theta - L = \tfrac{1}{2}(\nu^2 - \kappa^2)U_\theta^2 + V(U) \\ &= \tfrac{1}{2}(\nu^2 - \kappa^2)^{-1}\Pi^2 + V(U). \end{aligned}$$

The integral $H = A$ is solved as

$$\Pi = 2^{1/2}(\nu^2 - \kappa^2)^{1/2}[A - V(U)]^{1/2},$$

and

$$\begin{aligned} \mathscr{L} &= \frac{1}{2\pi}\oint \Pi \, dU - A \\ &= \frac{1}{2\pi}[2(\nu^2 - \kappa^2)]^{1/2}\oint [A - V(U)]^{1/2}\, dU - A, \end{aligned}$$

in agreement with (20).

In the linear case, $V(U) = \tfrac{1}{2}U^2$, the integral in $\mathscr{L}$ can be evaluated explicitly and we have

$$\mathscr{L} = [(\nu^2 - \kappa^2)^{1/2} - 1]A. \tag{107}$$

From the energy equation, A is proportional to a^2, and this result conforms with the general result in (12).

(b) *Linear example.* To illustrate the remarks on linear problems at the end of § 7, we do the linear Klein–Gordon case directly. We have

$$\bar{L} = \frac{1}{2\pi}\int_0^{2\pi} \{\tfrac{1}{2}(\nu^2 - \kappa^2)U_\theta^2 - \tfrac{1}{2}U^2\}\, d\theta.$$

We substitute

$$U = a\cos\theta,$$

and deduce

$$\bar{L} = \mathscr{L}^* = \tfrac{1}{4}(\nu^2 - \kappa^2 - 1)a^2. \tag{108}$$

The dispersion relation is written in a different form in (107) and (108), but the final results are equivalent. In any linear problem, $\mathscr{L} = F(\omega, \kappa)a^2$, and the slowly varying equations are

$$\mathscr{L}_a \propto F(\omega, \kappa) = 0, \tag{109}$$

$$\frac{\partial}{\partial T}(F_\omega a^2) + \frac{\partial}{\partial X}(F_\kappa a^2) = 0, \tag{110}$$

$$\frac{\partial \kappa}{\partial T} + \frac{\partial \omega}{\partial X} = 0. \tag{111}$$

We suppose (109) is solved in the form $\omega = W(\kappa)$. Then $F(W, \kappa) = 0$ is an identity and it follows that

$$C(\kappa)F_\omega(W, \kappa) + F_\kappa(W, \kappa) = 0,$$

where $C(\kappa) = W'(\kappa)$ is the group velocity. Equations (110), (111) may then be written

$$\frac{\partial}{\partial T}\{a^2 f(\kappa)\} + \frac{\partial}{\partial X}\{a^2 f(\kappa)C(\kappa)\} = 0, \tag{112}$$

$$\frac{\partial \kappa}{\partial T} + C(\kappa)\frac{\partial \kappa}{\partial X} = 0, \tag{113}$$

where $f(\kappa) = F_\omega(W, \kappa)$. Equation (112) may be expanded to

$$f(\kappa)\left\{\frac{\partial a^2}{\partial T} + \frac{\partial (Ca^2)}{\partial X}\right\} + f'(\kappa)\left\{\frac{\partial \kappa}{\partial T} + C\frac{\partial \kappa}{\partial X}\right\} = 0.$$

The second term is zero by (113). Finally, therefore, the equations reduce to

$$\frac{\partial a^2}{\partial T} + \frac{\partial}{\partial X}(a^2 C) = 0, \qquad \frac{\partial \kappa}{\partial T} + C(\kappa)\frac{\partial \kappa}{\partial X} = 0.$$

This final form shows that different choices of $F(\omega, \kappa)$ will lead to the same final equations provided that they have the same solution $\omega = W(\kappa)$.

(c) *Boussinesq equations for water waves.* In Boussinesq's approximation for long water waves the Lagrangian is

$$L = \tfrac{1}{2}u_t^2 - \tfrac{1}{2}gu^2 - u(\phi_t + \tfrac{1}{2}\phi_x^2),$$

where u is the water height and ϕ is a velocity potential (see Whitham [**1965b**]). The notation u for the height is to conform with the general notation in this paper. To lowest order,

$$L = \tfrac{1}{2}\nu^2 U_\theta^2 - \tfrac{1}{2}gU^2 - U\{\gamma + \nu\Phi_\theta + \tfrac{1}{2}(\beta + \kappa\Phi_\theta)^2\}.$$

Introducing the Hamiltonian transformation, as in (95) etc., we have

$$\Pi_1 = \frac{\partial L}{\partial U_\theta} = \nu^2 U_\theta, \qquad \Pi_2 = \frac{\partial L}{\partial \Phi_\theta} = -U\{\nu + \kappa(\beta + \kappa\Phi_\theta)\},$$

$$H = \tfrac{1}{2}\nu^{-2}\Pi_1^2 - \tfrac{1}{2}\kappa^{-2}U^{-1}\{\Pi_2 + U(\nu + \beta\kappa)\}^2 + U(\gamma + \tfrac{1}{2}\beta^2) + \tfrac{1}{2}gU^2.$$

Since Φ is absent from H, $\Pi_2 = B(X, T)$. Then, solving $H = A(X, T)$ for Π_1, we have

$$\begin{aligned}\mathscr{L} &= \frac{1}{2\pi}\oint \Pi_1\, dU - A \\ &= \frac{1}{2\pi}\nu \oint \{\kappa^{-2}U^{-1}(B + U\nu + U\beta\kappa)^2 - U(2\gamma + \beta^2) - gU^2 + 2A\}^{1/2} \\ &\qquad \cdot dU - A.\end{aligned}$$

This is the result obtained previously (Whitham [**1965b**]) with slight changes in notation.

The variational equations are (104), (105).

(d) *Central orbits.* The methods developed here for waves have required both extensions and new versions of ideas and techniques in mechanics. It is interesting to return to some of the questions in mechanics from this different point of view. A typical problem arises in central orbits and the theory of their adiabatic invariants. The theory of this paper applies with the simplification that the dependent variables u_i are functions of t alone. The slow variations are introduced, now, by slow variations in time of the central force field.

If r and ϕ are the radius and polar angle, respectively, the Lagrangian for an orbit in a force field with potential $V(r, \varepsilon t)$ is

$$L = \tfrac{1}{2}\dot{r}^2 + \tfrac{1}{2}r^2\dot{\phi}^2 - V(r, \varepsilon t).$$

The Euler equations are

$$\ddot{r} - r\dot{\phi}^2 + V_r(r, \varepsilon t) = 0, \qquad \frac{d}{dt}(r^2\dot{\phi}) = 0.$$

For $\varepsilon = 0$, the solution may be written

$$r = R(\theta), \qquad \theta = \nu t, \qquad \phi = \gamma t + \Phi(\theta),$$

where R and Φ are periodic functions of θ with period 2π. In this unperturbed case, ν and γ are constant. The period of r in time is $2\pi/\nu$. In that period, ϕ increases by $2\pi\gamma/\nu$; the orbit is closed if $\gamma = \nu$.

For the perturbed case $\varepsilon \neq 0$, the solution is generalized to the slowly varying form

$$r = R(\theta, T), \qquad T = \varepsilon t, \qquad \theta = \varepsilon^{-1}\Theta(T), \qquad \nu = \Theta_T,$$

$$\phi = \psi + \Phi(\theta, T), \qquad \psi = \varepsilon^{-1}\Psi(T), \qquad \gamma = \Psi_T.$$

To lowest order the Lagrangian is

$$L = \tfrac{1}{2}\nu^2 R_\theta^2 + \tfrac{1}{2}R^2(\gamma + \nu\Phi_\theta)^2 - V(R, T).$$

The "Hamiltonian transformation" is

$$\Pi_1 = \partial L/\partial R_\theta = \nu^2 R_\theta, \qquad \Pi_2 = \partial L/\partial \Phi_\theta = \nu R^2(\gamma + \nu\Phi_\theta),$$

$$H = \tfrac{1}{2}\nu^{-2}\Pi_1^2 + \tfrac{1}{2}\nu^{-2}R^{-2}\Pi_2^2 - \nu^{-1}\gamma\Pi_2 + V(R, T).$$

(This "Hamiltonian" differs by the term $\nu^{-1}\gamma\Pi_2$ from the usual one.) As before, Π_2 and H are integrals as regards the θ dependence. It is convenient to introduce corresponding parameters $M(T)$, $E(T)$, by

$$\Pi_2 = \nu M, \qquad H = E - \gamma M,$$

so that M and E are the angular momentum and energy, respectively. These are solved for Π_1 and Π_2 to give

$$\Pi_1 = \nu\{2E - M^2R^{-2} - 2V(R, T)\}^{1/2}, \qquad \Pi_2 = \nu M.$$

The average Lagrangian is

$$\bar{L} = \frac{1}{2\pi}\int_0^{2\pi} \{\Pi_1 R_\theta + \Pi_2\Phi_\theta - H\}\, d\theta.$$

Therefore

$$\mathscr{L} = \frac{\nu}{2\pi}\oint \{2E - M^2R^{-2} - 2V(R, T)\}^{1/2}\, dR + \gamma M - E.$$

The variations with respect to Θ and Ψ give

$$\frac{d}{dT}\mathscr{L}_\nu = 0, \qquad \frac{d}{dT}\mathscr{L}_\gamma = 0.$$

Hence,

$$I_1 = \mathscr{L}_\nu = \frac{1}{2\pi}\oint \{2E - M^2R^{-2} - 2V(R, T)\}^{1/2}\, dR,$$

$$I_2 = \mathscr{L}_\gamma = M$$

are the adiabatic invariants which remain constant for slow changes in the force field. The second shows that the angular momentum remains constant, as it must in a central force field. The first then determines the slow changes in the energy E.

The variations of $\mathscr{L}$ with respect to E and M give

$$\mathscr{L}_E = \nu \partial I_1/\partial E - 1 = 0, \qquad \mathscr{L}_M = \nu\partial I_1/\partial M + \gamma = 0.$$

These determine the frequencies ν and γ.

A convenient alternative form is to introduce I_1 and I_2 as parameters in place of E and M. Then

$$\mathscr{L} = \nu I_1 + \gamma I_2 - E(I_1, I_2),$$

and the variational equations are

$$\frac{dI_1}{dT} = 0, \qquad \frac{dI_2}{dT} = 0,$$

$$\nu = \frac{\partial E}{\partial I_1}, \qquad \gamma = \frac{\partial E}{\partial I_2}.$$

These are the standard results of the theory of adiabatic invariants, but they are usually obtained by other methods. A good account is given by Landau and Lifshitz [**1960**].

For the case of an inverse square force, the potential energy is $V = -\alpha(T)/R$. For this case the integral in $\mathscr{L}$ can be evaluated explicitly (most simply by contour integration) and we have

$$\mathscr{L} = \nu\left(\frac{\alpha}{[2|E|]^{1/2}} - M\right) + \gamma M - E.$$

The adiabatic invariants are

$$I_1 = \frac{\alpha}{[2|E|]^{1/2}} - M, \qquad I_2 = M.$$

The frequency equations $\mathscr{L}_E = \mathscr{L}_M = 0$ give

$$\nu = \gamma = (-2E)^{3/2}/\alpha.$$

The energy may be expressed in terms of I_1, I_2 by

$$E = -\frac{\alpha^2}{2(I_1 + I_2)^2}.$$

Since the frequencies are equal, the orbit is closed. As regards the general theory, the equal frequencies make this a degenerate case.

References

1968. J. D. Cole, *Perturbation methods in applied mathematics*, Blaisdell, Waltham, Mass., 1968. MR**39** #7841.

1959. G. E. Kuzmak, *Asymptotic solutions of nonlinear second order differential equations with variable coefficients*, Prikl. Mat. Meh. **23** (1959), 515–526 = J. Appl. Math. Mech. **23** (1959), 730–744. MR**22** #807.

1960. L. D. Landau and E. M. Lifshitz, *Mechanics*, Fizmatgiz, Moscow, 1958; English transl., Course of Theoretical Physics, vol. 1, Pergamon Press, Oxford; Addison-Wesley, Reading, Mass., 1960. MR**21** #985; MR**22** #11796.

1966. J. C. Luke, *A perturbation method for nonlinear dispersive wave problems*, Proc. Roy. Soc. Ser. A. **292** (1966), 403–412. MR**33** #3491.

1965a. G. B. Whitham, *Non-linear dispersive waves*, Proc. Roy. Soc. Ser. A. **283** (1965), 238–261. MR**31** #996.

1965b. ———, *A general approach to linear and non-linear dispersive waves using a Lagrangian*, J. Fluid Mech. **22** (1965), 273–283. MR**31** #6459.

1967a. ———, *Non-linear dispersion of water waves*, J. Fluid Mech. **27** (1967), 399–412. MR**34** #8711.

1967b. ———, Proc. Roy. Soc. Ser. A. **299** (1967), 6.

California Institute of Technology

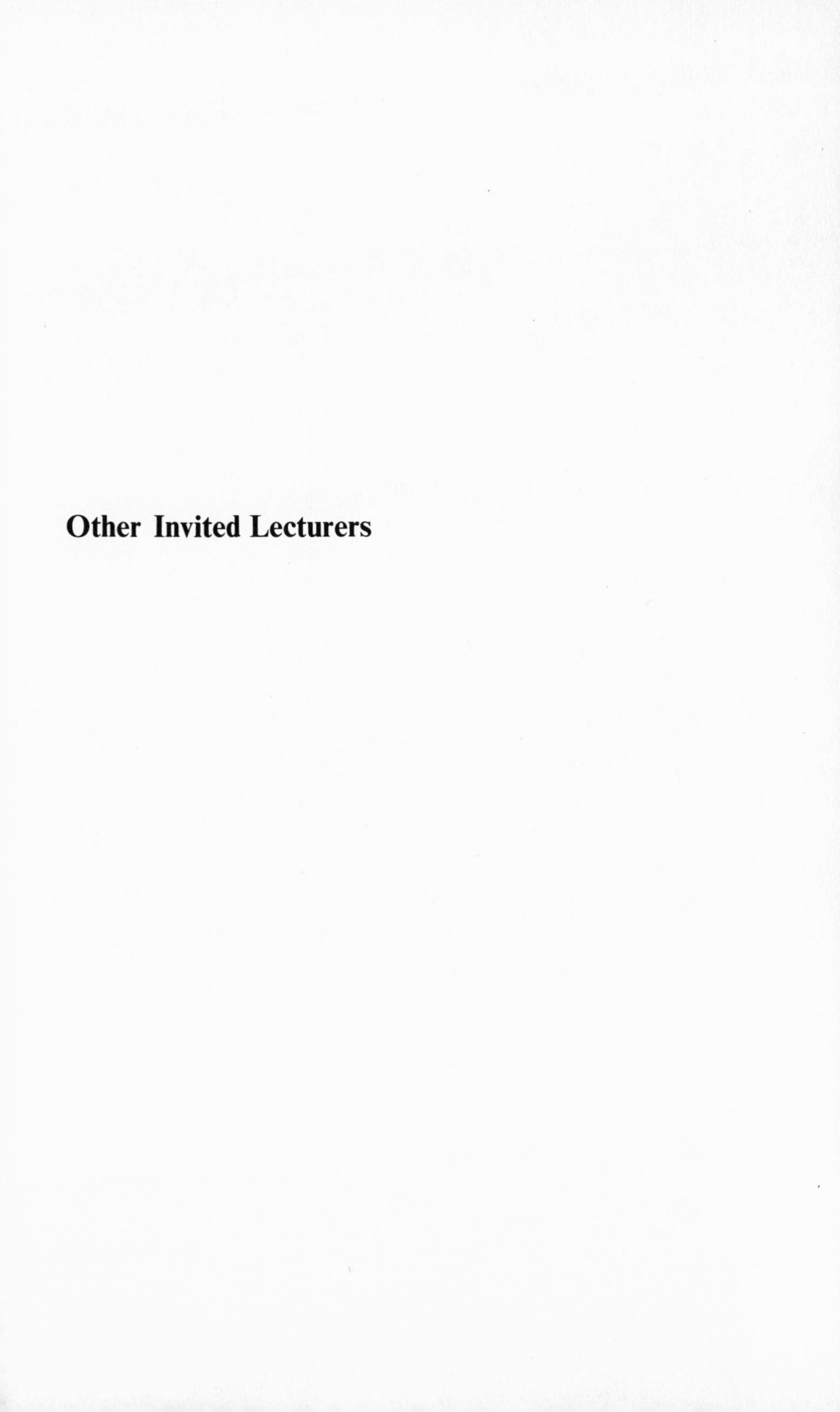

Other Invited Lecturers

Lectures in Applied Mathematics
Volume 15, 1974

Nonlinear Dispersive Waves and Multi-Phase Modes

Mark J. Ablowitz

In recent years there has been a renewed interest in the study of nonlinear wave motion. Probably the most familiar approach is the theory of weak nonlinear interactions. The general idea is to perturb known linear solutions and to study basic changes in the system due to the nonlinear couplings. While such weak interaction ideas adequately explain some observed phenomena, the method is restricted to situations in which the nonlinearity is extremely mild.

Subsequently, some notable advances have been made in fully nonlinear problems. In this regard we have Whitham's theory for the slow dispersion of fully nonlinear wavetrains ([**1**], [**2**], [**3**], [**4**], [**5**]) and the exact solution of the Korteweg–de Vries equation [**6**]. In this note I will concentrate on aspects of the former. Whitham's approach is to examine first those nonlinear problems which admit perfect uniform periodic wave solutions with constant frequency ω, wavenumber k, and amplitude E with an amplitude dependent dispersion relation $\omega = \omega(k, E)$. One can discuss the evolution of nonuniform or slowly varying wavetrains by deriving equations which govern the modulation of the quantities ω, k, E. The method can be formulated as a WKB procedure for nonlinear partial differential equations [**7**].

First, we examine a simplified model problem which allows us to investigate fundamentals of the theory and at the same time develop the tools necessary to generalize to fully nonlinear interacting waves. The latter analysis is based upon a knowledge of certain multiply periodic uniform

---

AMS (MOS) subject classifications (1970). Primary 35B25.

wave solutions which we refer to as multi-phase modes. The methods are general and results from other theories can be recovered as special cases ([**8**], [**9**]). The following complex nonlinear Klein–Gordon equation,

$$u_{tt} - u_{xx} + V'(uu^*)u = 0, \tag{1}$$

provides a representative model. Here u^* is the complex conjugate of u and $V'(uu^*)$ is any "reasonable" function of its argument. A uniform periodic wave solution is

$$u = E\, e^{i(kx - \omega t)} \tag{2}$$

so long as

$$\omega^2 - k^2 = V'(E^2). \tag{3}$$

We can find more general solutions to (1) by allowing ω, k and E to vary slowly. The mathematical procedure is to make use of the multiple scaling ideas [**7**]. Essentially, one looks for a solution of the problems as a function of a rapidly varying scale, in this case the phase of the wave, and slower scales governing the nonuniformities. Thus,

$$\begin{aligned} u &= u(\theta, X, T; \varepsilon), \\ \theta_t &= -\omega(X, T), \qquad \theta_x = k(X, T), \\ X &= \varepsilon x, \qquad\qquad T = \varepsilon t. \end{aligned} \tag{4}$$

Here ε is a small parameter. Using the definitions (4), equation (1) is found to be [**9**]

$$\begin{aligned} (\omega^2 - k^2)u_{\theta\theta} + V'(uu^*)u - \varepsilon[(\omega_T + k_X)u_\theta + 2(\omega u_{\theta T} + k u_{\theta X})] \\ + \varepsilon^2[u_{TT} - u_{XX}] = 0. \end{aligned} \tag{5}$$

Thus (1) a two-variable nonlinear problem is transformed into a three-variable nonlinear problem, and it is by no means clear that the situation is simplified. However, the limit $\varepsilon \ll 1$ allows an approximate solution to be obtained.

Indeed by expanding

$$\begin{aligned} u &= f + \varepsilon u^{(1)} + \varepsilon^2 u^{(2)} + \cdots, \\ \omega^2 - k^2 &= g + \varepsilon m^{(1)} + \varepsilon^2 m^{(2)} + \cdots, \\ \omega_T + k_X &= G + \varepsilon M^{(1)} + \varepsilon^2 M^{(2)} + \cdots, \\ k_T + \omega_X &= 0, \end{aligned} \tag{6}$$

and requiring $u(\theta)$ to be periodic with period 2π, the complete set of slow modulation equations can be deduced. They are

$$
\begin{aligned}
&\omega^2 - k^2 = V'(E^2) + \varepsilon^2((E_{TT} - E_{XX})/E), \\
(7) \qquad &(\omega E^2)_T + (kE^2)_X = 0, \\
&k_T + \omega_X = 0.
\end{aligned}
$$

The solution $u(\theta, X, T; \varepsilon)$ is given by

$$
(8) \qquad u = E\, e^{i\theta}
$$

where E and θ are found from solving system (7). Thus the analysis results in a truncated system and allows reasonable discussion of normally difficult issues.

Insofar as solutions to (7) are concerned, one usually solves by consistently neglecting the ε^2 contribution. The reduced equations are often referred to as Whitham's equations. They comprise a system of first order nonlinear partial differential equations. Corresponding to this system is a set of characteristics which may be either real or complex. If the characteristics are real, we term the system hyperbolic. For each real characteristic there is a distinct velocity which in the linear limit ($V'(E^2) = 1$) merge together to give the familiar group velocity.

Furthermore, since the system is hyperbolic, we have to expect that the solutions of the reduced system may become multi-valued after a sufficient time. However, in these situations we must keep the higher terms in the series since they become as important as the leading order terms at a shock front where derivatives are quite large. For (7) when the ε^2 term is retained, no smooth steady shock transition is possible [**9**]. Effectively the ε^2 term disperses the shock front. In this context it should be mentioned that the reduced equation can be hyperbolic yet become singular in a variety of ways, other than forming discontinuities. Recently it has been shown that the reduced system corresponding to the Korteweg–de Vries equation develops singular behavior whenever solitons are present [**10**].

In other cases the characteristics of the reduced system can be complex indicating that the equations are elliptic. Elliptic initial value problems are in a sense unstable; hence the wave solution (8) will not retain its coherence over a long time. This is similar to the instability first found by Benjamin (in the Stokes water wave solution [**11**]). However, instability can manifest itself in other ways as well. The reduced system can be hyperbolic yet the solution (8) can still be unstable. In these cases the ε^2 term in (7) has a destabilizing influence. The indication is that the system is stable with respect to long wave perturbations but unstable with respect to short scales. Thus one must in general verify the stability of the nonlinear wave for order one times. This condition is *not* inherent in the hyperbolicity of the reduced system.

While Whitham's theory and its ramifications are quite interesting and suggestive, so is its extension to far more complicated situations. Specifically we are concerned with the evolution of fully nonlinear interacting waves which have a multiply periodic structure. The ideas are similar to those presented above except that the basic wavetrains now will be allowed to consist of many different phases.

The essential features of the method are illustrated by considering two-phase modes and the model problem, a real nonlinear Klein–Gordon equation,

$$u_{tt} - u_{xx} + V'(u) = 0. \tag{9}$$

Here one looks for a solution of (9) as a function of two rapidly varying scales, the phases of the wave and slower scales governing the nonuniformities. Thus,

$$\begin{aligned} &u = u(\theta_1, \theta_2, X, T; \varepsilon), \\ &\theta_{it} = -\omega_i(X, T), \qquad \theta_{ix} = k_i(X, T) \qquad (i = 1, 2), \\ &X = \varepsilon x, \qquad\qquad T = \varepsilon t, \end{aligned} \tag{10}$$

and again $\varepsilon \ll 1$.

Using (10) equation (9) is put into the form

$$\begin{aligned} (\omega_1^2 - k_1^2)u_{\theta_1\theta_1} &+ (\omega_2^2 - k_2^2)u_{\theta_2\theta_2} + 2(\omega_1\omega_2 - k_1k_2)u_{\theta_1\theta_2} + V'(u) \\ &- \varepsilon[(\omega_{1T} + k_{1X})u_{\theta_1} + (\omega_{2T} + k_{2X})u_{\theta_2} \\ &\quad + 2(\omega_1 u_{\theta_1 X} + k_1 u_{\theta_1 T}) + 2(\omega_2 u_{\theta_2 X} + k_2 u_{\theta_2 T})] \\ &+ \varepsilon^2(u_{XX} - u_{TT}) = 0. \end{aligned} \tag{11}$$

The perturbation procedure is to expand

$$\begin{aligned} u &= f + \varepsilon u^{(1)} + \varepsilon^2 u^{(2)} + \cdots, \\ \omega_i^2 - k_i^2 &= g_i + \varepsilon M_i^{(1)} + \varepsilon^2 M_i^{(2)} + \cdots, \\ \omega_{iT} + k_{iX} &= G_i + \varepsilon M_i^{(1)} + \varepsilon^2 M_i^{(2)} + \cdots, \\ k_{iT} + \omega_{iX} &= 0, \qquad i = 1, 2. \end{aligned} \tag{12}$$

With the definition $\lambda = \omega_1\omega_2 - k_1k_2$, substitution of (12) into (11) yields the following sequence of problems:

$$g_1 f_{\theta_1\theta_1} + g_2 f_{\theta_2\theta_2} + 2\lambda f_{\theta_1\theta_2} + V'(f) = 0, \tag{13}$$

$$Lu^{(n)} = g_1 u^{(n)}_{\theta_1\theta_1} + g_2 u^{(n)}_{\theta_2\theta_2} + 2\lambda u^{(n)}_{\theta_1\theta_2} + V''(f)u^{(n)} = F^{(n)}. \tag{14}$$

(13) is the basic nonlinear problem for the two-phase mode $f(\theta_1, \theta_2)$, and

(14) is the linearized problem obtained by perturbing f. $F^{(n)}$ depends on $f, u^{(1)}, u^{(2)}, \ldots, u^{(n-1)}$.

Given a doubly periodic function f satisfying (13) it is necessary to choose the arbitrary functions $m_i^{(j)}$, $M_i^{(j)}$ appropriately to ensure that each $u^{(n)}$ is multiply periodic in θ_1, θ_2, period 2π. Using the appropriate secular conditions the entire expansion can be obtained. For details the reader is referred to [**8**].

The analogous equations to that found by Whitham are

$$\begin{aligned}
&\int_0^{2\pi}\int_0^{2\pi} [(\omega_{1T} + k_{1X})f_{\theta_1}^2 + (\omega_{2T} + k_{2X})f_{\theta_1}f_{\theta_2} \\
&\qquad + 2(\omega_1 f_{\theta_1 T} + k_1 f_{\theta_1 X})f_{\theta_1} + 2(\omega_2 f_{\theta_2 T} + k_2 f_{\theta_2 X})f_{\theta_1}]\, d\theta_1\, d\theta_2 = 0, \\
&\int_0^{2\pi}\int_0^{2\pi} [(\omega_{2T} + k_{2X})f_{\theta_2}^2 + (\omega_{1T} + k_{1X})f_{\theta_1}f_{\theta_2} \\
&\qquad + 2(\omega_1 f_{\theta_1 T} + k_1 f_{\theta_1 X})f_{\theta_2} + 2(\omega_2 f_{\theta_2 T} + k_2 f_{\theta_2 X})f_{\theta_2}]\, d\theta_1\, d\theta_2 = 0, \\
&\qquad k_{iT} + \omega_{iX} = 0, \\
&\qquad \omega_i^2 - k_i^2 = g_i, \qquad i = 1, 2, \ldots,
\end{aligned} \tag{15}$$

where the g_i are defined in terms of the amplitude parameters, E_1, E_2,

$$\begin{aligned}
E_1 &= \frac{1}{2\pi}\int_0^{2\pi} \left(\frac{g_1}{2} f_{\theta_1}^2 - \frac{g_2}{2} f_{\theta_2}^2 + V(f)\right) d\theta_2, \\
E_2 &= \frac{1}{2\pi}\int_0^{2\pi} \left(\frac{g_2}{2} f_{\theta_2}^2 - \frac{g_1}{2} f_{\theta_1}^2 + V(f)\right) d\theta_1.
\end{aligned} \tag{16}$$

The system of equations (15) provides a set of first order coupled nonlinear equations which governs the evolution of the two-phase mode in a slowly varying background. Questions of stability, shocks, etc., can be discussed.

The above analysis is general and can be extended to modes with N phase variables. Previous theories can be recovered in special limits (i.e. single phase theory, weak nonlinearity, etc.).

While the ideas are conceptually appealing, a knowledge of the function f satisfying the fully nonlinear problem is necessary before the detailed evolution of the wave properties can be determined. Obviously this is not an easy question and in general an approximate numerical approach may be the most feasible [**12**].

REFERENCES

1. G. B. Whitham, Proc. Roy. Soc. Ser. A. **283** (1965), 238–261. MR**31** # 996.
2. ———, J. Fluid Mech. **22** (1965), 273–283. MR**31** # 6459.
3. ———, J. Fluid Mech. **27** (1967), 399–412. MR**34** # 8711.
4. ———, Proc. Roy. Soc. Ser. A. **299** (1967), 6.
5. ———, J. Fluid Mech. **44** (1970), 373–395. MR**42** #4858.
6. C. S. Gardner, J. M. Greene, M. D. Kruskal and R. M. Miura, Phys. Rev. Lett. **19** (1967), 1095.
7. J. C. Luke, Proc. Roy. Soc. Ser. A. **292** (1966), 403–412. MR**33** #3491.
8. M. J. Ablowitz and D. J. Benney, Studies in Appl. Math. **49** (1970), 225.
9. M. J. Ablowitz, Studies in Appl. Math. **50** (1971), 329.
10. G. Krishna, Ph.D. Thesis, Clarkson College.
11. T. B. Benjamin and J. E. Feir, J. Fluid Mech. **27** (1967), 417.
12. M. J. Ablowitz, Studies in Appl. Math. **51** (1972), 17.

Lectures in Applied Mathematics
Volume 15, 1974

A Brief Introduction to Ray Methods for Dispersive Hyperbolic Equations[1]

Norman Bleistein

Abstract. The ray method for dispersive hyperbolic equations is described via a specific example, i.e., application to the Klein–Gordon equation. Reference is made to other situations such as systems of partial differential equations and wave guide problems. A fairly extensive bibliography is included.

The ray method is a formal technique for obtaining asymptotic solutions to partial differential equations and systems. It was developed by J. B. Keller and his associates for the analysis of scattering problems in acoustics and electromagnetics ([**1**]–[**3**], [**8**]–[**10**], [**14**]–[**17**], [**19**]–[**25**]). Influenced by this work and also by some of G. B. Whitham's work ([**26**], [**28**]), R. M. Lewis initiated a systematic development of ray methods for dispersive equations and systems [**18**].

In this method, one starts by assuming a form of solution, substitutes into the given equation(s) and deduces a system of ordinary differential equations for the group lines (the *rays* of the ray method) and the propagation of the wave vector, frequency, phase and amplitude along the rays. This system is often referred to as the *ray equations* and *transport equation* (for amplitude propagation). The specific terminology of the method is an indication of the evolution of this method from geometrical optics.

AMS (*MOS*) *subject classifications* (1969). Primary 3506, 3509, 3516, 3531, 3553, 4150.

Key words and phrases. Ray method, dispersive, hyperbolic, characteristics, group velocity, rays, ray equations, transport equation, waves.

[1] Partially supported by the Office of Naval Research under contract N00014-67-A-0394-0005.

The use of the ray method makes unnecessary the analytical representation of exact solutions in order to derive their asymptotic expansions. Furthermore, the method avoids excursions through (at times obscure and even opaque) physics for the sake of deriving a *Lagrangian* for the given equation or system, out of which the ray equations and transport equation can also be derived; this is Whitham's method ([**27**], [**28**]). Finally, we note that one can obtain as many terms as desired of the asymptotic solution via the ray method.

In this brief introduction, I shall apply the ray method to the Klein–Gordon equation to show how the ray equations and transport equation are derived, but I shall also indicate at various places what might happen in a variety of alternative situations.

We begin by considering the equation

$$\nabla(c^2\nabla u) - u_{tt} - \lambda^2 b^2 u = 0. \tag{1}$$

Here, b and c are functions of $\boldsymbol{x} = (x_1, x_2, x_3)$ and t, and we seek an asymptotic solution for large values of λ. For the case of b and c constant, we could try a plane wave solution

$$u = A\exp\{i\lambda[\boldsymbol{k}\cdot\boldsymbol{x} - \omega t]\}, \qquad \boldsymbol{k} = (k_1, k_2, k_3), \tag{2}$$

and find that u satisfies (1) so long as

$$\omega = \pm h(k) = \pm(c^2k^2 + b^2)^{1/2}, \qquad k = |\boldsymbol{k}|. \tag{3}$$

Indeed, in this case we can write down the general solution to the free space initial value problem as a superposition of such plane waves,

$$u = \int \sum_{\omega = \pm h} A^{\pm}(\boldsymbol{k})\exp\{i\lambda[\boldsymbol{k}\cdot\boldsymbol{x} - \omega t]\}\, dk_1\, dk_2\, dk_3. \tag{4}$$

Returning to (1), we assume that

$$\begin{aligned} u &= \exp\{i\lambda\phi(\boldsymbol{x}, t)\}Z(\boldsymbol{x}, t; \lambda), \\ Z(\boldsymbol{x}, t; \lambda) &\sim \sum_{n=0}^{\infty} Z_n(\boldsymbol{x}, t)(i\lambda)^{-n+\alpha}. \end{aligned} \tag{5}$$

Such a form is well motivated by the asymptotic analysis of exact solutions, when they can be obtained. We substitute (5) into (1) and equate to zero the coefficient of each power of λ, obtaining

$$MZ_0 = [-c^2k^2 + \omega^2 - b^2]Z_0 = 0, \tag{6}$$

$$MZ_1 = -LZ_0 = -\{2c^2\boldsymbol{k}\cdot\nabla Z_0 - 2\omega Z_{0t} + Z_0[\nabla\cdot(c^2\nabla\phi) - \phi_{tt}]\}, \tag{7}$$

$$MZ_{n+1} + LZ_n = Z_{n-1tt} - \nabla\cdot(c^2\nabla Z_{n-1}). \tag{8}$$

Here

$$\boldsymbol{k} = \nabla\phi, \qquad \omega = -\phi_t. \tag{9}$$

To obtain a nontrivial result, we find from (6) that ω must satisfy one of the dispersion relations defined by (3), except that now that equation is a first order nonlinear partial differential equation for the phase function ϕ. When dealing with a system of equations, the coefficients $Z_0, Z_1, \ldots$ are vectors and M is a matrix. Dispersion relations then arise as conditions under which the matrix M is singular, and thus (6) has nontrivial solution vectors ([**17**], [**18**]). When studying propagation in a wave guide, (6) is replaced by a homogeneous boundary value problem involving a differential equation in as many independent variables as there are transverse dimensions in the wave guide ([**5**], [**13**]). Again, dispersion relations arise as conditions under which nontrivial solutions Z_0 of (6) exist.

We anticipate a propagating wave associated with each solution of (3). A total asymptotic solution u will then be made up of a superposition of such waves.

We propose to solve (3) by the method of characteristics [**7**]. In the context of this method we shall refer to the characteristic equations as the *ray equations*. They are

$$\begin{aligned}
\dot{\boldsymbol{x}} &= \boldsymbol{v}_g = \nabla_k\omega = c^2\boldsymbol{k}/\omega, \\
\dot{t} &= 1, \\
\dot{\phi} &= \boldsymbol{k}\cdot\dot{\boldsymbol{x}} - \omega\dot{t} = -b^2/\omega, \\
\dot{\boldsymbol{k}} &= -ck^2\nabla c/\omega - b\nabla b/\omega, \\
\dot{\omega} &= ck^2c_t/\omega + bb_t/\omega.
\end{aligned} \tag{10}$$

Here $(\dot{\ }) = d/ds_0$, with s_0 a parameter along the curves defined by the first two lines of this equation. From the second line here, we see that s_0 is just t, up to a translation. Thus it makes sense that $\dot{\boldsymbol{x}}$ is indeed a velocity, and we see from the first line that, in fact, $\dot{\boldsymbol{x}}$ is just the *group velocity*.

The appropriate problem for (10) is an initial value problem with *initial* here meaning at $s_0 = 0$, which is not necessarily the same as $t = 0$. To differentiate between the initial data for (10) and initial data for $t = 0$, we shall refer to the former as *ray data*. Let us suppose then that we are given ray data for (10), say, in parametric form with parameters $(s_1, s_2, s_3) = \boldsymbol{s}$:

$$\begin{aligned}
\boldsymbol{x} = \boldsymbol{x}_0(\boldsymbol{s}), \qquad t = t_0(\boldsymbol{s}), \qquad \phi = \phi_0(\boldsymbol{s}),& \\
\boldsymbol{k} = \boldsymbol{k}_0(\boldsymbol{s}), \qquad \omega = \omega_0(\boldsymbol{s}),& \qquad s_0 = 0.
\end{aligned} \tag{11}$$

A solution to (10), (11) provides a family of curves, the *rays* of the ray method, and describes the propagation of the wave vector, frequency and phase along the rays. From the form of the velocity on the rays we see that they are just the group lines. From (10) we also see that, if the coefficients b and c are time independent, then the frequency is *constant* on the rays, i.e., the frequency is independent of s_0 and thus is given by its initial value on the ray as determined from (11). Similarly, if b and c are $\boldsymbol{x}$ independent, then $\boldsymbol{k}$ is constant on rays.

We turn now to (7) the (first) transport equation and refer the reader to the literature for a formulation of this equation as an ordinary differential equation with respect to s_0 **[18]**. To quote the result, we must first introduce the jacobian

$$J = \partial(t, \boldsymbol{x})/\partial(s_0, \boldsymbol{s}). \tag{12}$$

Then one finds that

$$[Z_0^2 J\omega]^{\cdot} = 0 \tag{13}$$

with solution

$$Z_0 = f(\boldsymbol{s})|J\omega|^{-1/2}. \tag{14}$$

For the case in which the problem for u is an initial value problem, $t = s_0$ and the first row of the matrix for J is just $(1\ 0\ 0\ 0)$. Here then,

$$J = \partial(\boldsymbol{x})/\partial(\boldsymbol{s}). \tag{15}$$

For initial data with support at one point, we use the conoidal solution for ϕ in which the initial value of $\boldsymbol{k}$ is free, say $\boldsymbol{k} = \boldsymbol{\kappa}$ for $t = 0$. In this case, one may take $\boldsymbol{s} = \boldsymbol{\kappa}$, that is, the rays are "labelled" by their initial wave vector. In this case,

$$J = \partial(\boldsymbol{x})/\partial(\boldsymbol{\kappa}). \tag{16}$$

This is the case which arises when the initial data in dimensional variables have compact support and large λ represents "large" time.

We remark also that (7) can be recast in the more familiar "conservation" form

$$(Z_0^2\omega)_t + \nabla \cdot (Z_0^2\omega \boldsymbol{v}_g) = 0. \tag{17}$$

In the formalisms arising from a system of equations and/or arising from propagation in a wave guide, making M singular in (6) allows one to solve for Z_0 to within one or more multiplicative constants. Furthermore, in these cases, (7) will have a solution for Z_1 only if the inhomogeneous term in that equation satisfies orthogonality conditions with respect to the solutions of the adjoint of the homogeneous problem (6). These orthogonality conditions take the form of transport equations and serve to determine the multiplicative constants mentioned above. Of all the computations associated with those manipulations associated with recasting these orthogonality conditions in the form of conservation laws, either in the form (13) or the equivalent form (17), these are often the most elusive. Even when this is not possible, one can still expect to write these transport equations as ordinary differential equations along the rays.

Returning now to (8), the higher order transport equations, we note that each of these equations is an inhomogeneous ordinary differential equation along the rays. This follows, since $M = 0$ and the operator L is equivalent to the ordinary differential operator implicit in (13). Thus, one solves this system recursively for the Z_n's when ray data for them are given.

To complete our formalism, we shall demonstrate a few ways in which ray data are determined from a problem for u. As a first example, we consider the case in which we are given an initial value problem for u and the data are already of the form of the assumed solution. That is, let us suppose that we are given the initial data

$$\begin{aligned} u(\boldsymbol{x}, 0) &= U(\boldsymbol{x}; \lambda) \exp\{i\lambda\psi(\boldsymbol{x})\}, \qquad u_t(\boldsymbol{x}, 0) = 0, \\ U(\boldsymbol{x}; \lambda) &\sim \sum_{n=0}^{\infty} U_n(\boldsymbol{x})(i\lambda)^{-n}. \end{aligned} \tag{18}$$

We assume that u is given as a sum of one or more waves of the form (5) and conclude first from the initial data that for each such wave

$$\phi(\boldsymbol{x}, 0) = \psi(\boldsymbol{x}). \tag{19}$$

By differentiating this equation with respect to each x_i, $i = 1, 2, 3$, we find that

$$\boldsymbol{k}(\boldsymbol{x}, 0) = \nabla\psi. \tag{20}$$

Now, from the dispersion equation (3) we conclude that there must be two waves of the form (5), distinguished at first by the ray data for the frequency

$$\omega(\boldsymbol{x}, 0) = \pm(c^2(\nabla\psi)^2 + b^2)^{1/2}. \tag{21}$$

Thus we write

$$u(\boldsymbol{x}, t; \lambda) \sim \sum_{n=0, \pm}^{\infty} Z_n^{\pm}(\boldsymbol{x}, t)(i\lambda)^{-n} \exp\{i\lambda\phi^{\pm}(\boldsymbol{x}, t)\}. \tag{22}$$

We differentiate this expansion with respect to t and then evaluate both u and u_t at $t = 0$ using (18). This leads to the following system of equations for the initial values of the Z_n's:

$$\sum_{\pm} Z_0^{\pm}(\boldsymbol{x}, 0) = U_0(\boldsymbol{x}), \qquad \sum_{\pm} \pm (c^2(\nabla\psi)^2 + b^2)^{1/2} Z_0^{\pm} = 0,$$

$$(23) \qquad \sum_{\pm} Z_n^{\pm}(\boldsymbol{x}, 0) = U_n(\boldsymbol{x}), \qquad \sum_{\pm} \pm (c^2(\nabla\psi)^2 + b^2)^{1/2} Z_n^{\pm} + Z_{n-1t}^{\pm} = 0,$$

$$n = 1, 2, \ldots .$$

We can solve explicitly for $Z_0^{\pm}(\boldsymbol{x}, 0)$ here. The ray data for each subsequent coefficient are then determined only after the problem for the previous coefficient has been solved, or at least this predecessor's partial derivative with respect to t is determined initially. At least for the leading order we have the result

$$(24) \qquad Z_0^{+}(\boldsymbol{x}, 0) = Z_0^{-}(\boldsymbol{x}, 0) = U_0(\boldsymbol{x})/2.$$

To recast these results in the form (11), we first set

$$(25) \qquad \boldsymbol{x}_0^{\pm}(\boldsymbol{s}) = \boldsymbol{s}.$$

That is, the parameter $\boldsymbol{s}$ is the point of initiation of each ray. We then complete the list in (11) in the following way:

$$(26) \qquad \begin{aligned} t_0^{\pm} &= 0, & \phi_0^{\pm} &= \psi(\boldsymbol{s}), \\ \boldsymbol{k}_0^{\pm} &= \nabla\psi(\boldsymbol{s}), & \omega_0^{\pm} &= \pm(c^2(\boldsymbol{s}, 0)k_0^2 + b^2(\boldsymbol{s}, 0))^{1/2}. \end{aligned}$$

Finally, we note that for this choice of $\boldsymbol{s}$, the jacobian J defined by (15) is initially equal to one, for either family of rays. Thus in (14),

$$(27) \qquad f^{\pm}(\boldsymbol{s}) = U_0(\boldsymbol{s})|\omega_0(\boldsymbol{s})|^{1/2}/2.$$

This completes the formulation of ray data for this case.

As a second example, we consider an initial value problem with initial data

$$(28) \qquad u(\boldsymbol{x}, 0) = p(\lambda(\boldsymbol{x} - \boldsymbol{\xi})), \qquad u_t(\boldsymbol{x}, 0) = i\lambda q(\lambda(\boldsymbol{x} - \boldsymbol{\xi})).$$

Here, the functions p and q are assumed to have "compact support" in their arguments so that as $\lambda \to \infty$ the support of the initial data shrinks to the point $\boldsymbol{\xi}$. This form of data is the generalization of multi-pole data.

To determine the ray data for this problem we must resort to an indirect method. Firstly, we assume that all the rays emanate from the point $\boldsymbol{\xi}$ and therefore we can immediately identify two of the functions in (11), namely,

$$(29) \qquad \boldsymbol{x}_0 = \boldsymbol{\xi}, \qquad t_0 = 0.$$

Next we argue that the ray data can only depend on local properties of the medium near $(\boldsymbol{\xi}, 0)$ and, to leading order at least, should depend on the values of b and c at that point only. Thus we expect that the function f

appearing in (14) should be the same for the given problem as it is for the problem in which b and c are replaced by their values $b(\xi, 0)$ and $c(\xi, 0)$, respectively.

For the constant coefficient case, we can solve the system of equations (10) and (13) explicitly, modulo the constants we seek. The result is given parametrically with parameters $\boldsymbol{\kappa} = (\kappa_1, \kappa_2, \kappa_3)$:

$$
(30)\qquad
\begin{aligned}
u &\sim \sum_{\pm} Z_0^{\pm}(\boldsymbol{x}, t)\lambda^{\alpha} \exp\{i\lambda\phi^{\pm}(\boldsymbol{x}, t)\},\\
\phi^{\pm} &= \phi_0^{\pm} + \boldsymbol{\kappa}\cdot(\boldsymbol{x} - \boldsymbol{\xi}) \mp \omega t,\\
\omega &= (c^2\kappa^2 + b^2)^{1/2},\\
\boldsymbol{x} &= \boldsymbol{\xi} \pm \nabla_{\kappa}\omega t = \boldsymbol{\xi} \pm (c^2\boldsymbol{\kappa}/\omega)t,\\
\boldsymbol{k} &= \boldsymbol{\kappa},\\
J &= \frac{\partial(\boldsymbol{x})}{\partial(\boldsymbol{\kappa})} = \left|\det\left(\frac{\partial^2\omega}{\partial\boldsymbol{\kappa}_i\,\partial\boldsymbol{\kappa}_j}t\right)\right|,\\
Z_0^{\pm} &= f^{\pm}(\boldsymbol{\kappa})/|\omega J|^{1/2}.
\end{aligned}
$$

In this equation we have allowed for the possibility of a multiplicative factor λ^{α}. We have also made explicit use of the fact that $s_0 = t$. We have departed from our previous use of ω by setting it equal to the positive square root here and then explicitly inserting the $\pm$ signs where they should appear. From (10), with b and c constant, we see that ω and $\boldsymbol{k}$ are constant on rays and thus given by their initial values. The solutions $\phi^{\pm}$ are the conoidal solutions, i.e., solutions of the dispersion equations all of whose rays emanate from one point, in this case the point $(\boldsymbol{\xi}, 0)$. To find $\phi^{\pm}$ and $Z_0^{\pm}$ for a given $(\boldsymbol{x}, t)$, one first solves the ray equation for $\boldsymbol{\kappa}(\boldsymbol{x}, t)$ and substitutes this value for $\boldsymbol{\kappa}$ wherever it appears. For each such solution $\boldsymbol{\kappa}$ one obtains a contribution to the solution u and one must sum all of these to find the total asymptotic solution for the given $(\boldsymbol{x}, t)$.

Actually, in this specific example, one could obtain an explicit expression $\boldsymbol{\kappa}(\boldsymbol{x}, t)$ and eliminate the parameters in the solution. However, that elimination obscures the locally planar form of the phase which is apparent in (30), and furthermore that elimination does not help us with our primary objective here, which is to determine $\phi_0^{\pm}$ and $f^{\pm}$. To this end, we use the representation (4) of the solution for constant b and c. The method of multi-dimensional stationary phase can be applied to that result to obtain the leading term of the asymptotic expansion of u, with $A^{\pm}(\boldsymbol{k})$ given in terms of the fourier transform of the initial data. This result turns out to be of exactly the same form as (30) except that the unknown constants of (30) are explicit in this latter result. When these operations are carried out and the indicated comparison is made, we find that

$$(31)\quad f^{\pm} = \pm \frac{\exp[\mp i\sigma\pi/4]}{2\omega} \int_{-\infty}^{\infty} [p(\boldsymbol{\eta}) \mp \omega q(\boldsymbol{\eta})] \exp[-i\boldsymbol{\kappa} \cdot \boldsymbol{\eta}]\, d\eta_1\, d\eta_2\, d\eta_3,$$

$$\phi_0^{\pm} = 0.$$

Here, σ is the difference between the number of positive and negative eigenvalues of the matrix of second derivatives. We now take these values given in (31) to be the appropriate ray data for the problem with variable coefficients as well, so long as b and c are properly interpreted as discussed above. This completes the discussion of this example.

When a wave is incident on a boundary surface (i.e., when the rays associated with a solution of the form (5) impinge on a surface) we expect that they give rise to another wave, the reflected wave. The surface is described in space-time by equations of the form

$$(32)\qquad \boldsymbol{x} = \boldsymbol{x}_0(s_1, s_2), \qquad t = t_0(\boldsymbol{s}) = s_3,$$

and these serve as ray data for $\boldsymbol{x}$ and t for the reflected wave, when we assume that it, too, has the form (5). Let us assume that the total field is of the form

$$(33)\qquad u = u^I + u^R$$

and must satisfy a boundary condition of the form

$$(34)\qquad \partial u/\partial N - i\lambda\zeta u = 0.$$

Here, $\partial/\partial N$ denotes the normal derivative to the boundary surface while the superscripts I and R connote "incident" and "reflected" waves, respectively, and we shall use the same superscripts below. It is assumed that the incident wave is known and we seek the remainder of the ray data for the reflected wave in terms of this known wave and the boundary condition.

Firstly, we require that the phases of the two waves be equal,

$$(35)\qquad \phi^R(\boldsymbol{x}_0(s_1, s_2), s_3) = \phi^I(\boldsymbol{x}_0(s_1, s_2), s_3) = \phi_0^R(\boldsymbol{s}).$$

Differentiation of this equation with respect to each of the parameters then leads to the results

$$(36)\qquad \begin{aligned} \boldsymbol{k}_T^R &= \boldsymbol{k}_T^I(\boldsymbol{x}_0(s_1, s_2), s_3) = \boldsymbol{k}_0^R(\boldsymbol{s}), \\ \omega^R &= \omega^I(\boldsymbol{x}_0(s_1, s_2), s_3) = \omega_0^R(\boldsymbol{s}). \end{aligned}$$

Here the subscript T denotes the tangential component of the wave vector. To find the normal component of $\boldsymbol{k}^R$ we use the dispersion equation. For the specific case (3) we choose the root

$$(37)\qquad k_N^R = -k_N^I(\boldsymbol{x}_0(s_1, s_2), s_3) = k_{N_0}^R(\boldsymbol{s}),$$

since the alternative choice would lead to $u^R = -u^I$ and $u = 0$, which we reject as a meaningless result. To determine ray data for the amplitude we use the boundary condition and find, for example for Z_0^R,

$$Z_0^R(\boldsymbol{x}_0(s_1, s_2), s_3) = \frac{k_N^I - \zeta}{k_N^I + \zeta} Z_0^I(\boldsymbol{x}_0(s_1, s_2), s_3). \tag{38}$$

The ray method has proved to be particularly useful for the analysis of diffraction phenomena. Again, a major task is to find ray data for diffracted waves. This can be done via indirect methods, using canonical problems, much as in the second example above. Typically, the ray method for the reflected wave breaks down in the neighborhood of points on the boundary surface which give rise to diffracted rays. For example, the ray data might have a discontinuity along some curve (edge diffraction) or the data might be "characteristic"—$J = 0$ (smooth body diffraction).

Such breakdowns can also occur away from boundary surfaces. For example, they occur on envelopes of the ray field, such as smooth caustics, cusped caustics and foci, where $J = 0$, and also along "shadow boundaries" of the ray field, where the asymptotic solution is discontinuous.

Recently, many of these anomalies have been studied by resorting to so-called uniform ray methods. Here, one assumes a more general wave-like function than just the exponential appearing in (5). Of course, the specific choice of function is motivated by study of canonical problems for the specific anomaly at hand. Typically, away from the anomaly, the special function is replaced by its asymptotic expansion to reproduce the form (5), while right at the anomaly, the special function remains finite, thus avoiding the breakdown inherent in the continuation of (5) to the anomaly. Many examples of this method are contained in the references.

REFERENCES

1. D. S. Ahluwahlia, *Uniform asymptotic theory of diffraction by the edge of a three dimensional body*, SIAM J. Appl. Math. **18** (1970), 287–301. MR**41** #6471.

2. D. S. Ahluwahlia, R. M. Lewis and J. Boersma, *Uniform asymptotic theory of diffraction by a plane screen*, SIAM J. Appl. Math. **16** (1960), 783–807.

3. N. Bleistein and R. M. Lewis, *Space-time diffraction for dispersive hyperbolic equations*, SIAM J. Appl. Math. **14** (1966), 1454–1470. MR**37** #607.

4. N. Bleistein, *Diffraction in the asymptotic solution of a dispersive hyperbolic equation*, Arch. Rational Mech. Anal. **31** (1968/69), 214–227. MR**38** #2422.

5. N. Bleistein, J. K. Cohen and D. Hector, *Ray method expansions for surface and internal waves in inhomogeneous oceans of variable depth*, Studies Appl. Math. **L1** (1972), 121–137.

6. N. Bleistein, *Ray methods for partial differential equations*, Lecture Notes, North British Differential Equations Sympos., University of Dundee, 1972.

7. R. Courant and D. Hilbert, *Methods of mathematical physics*. Vol. II: *Partial differential equations*, Interscience, New York, 1962. MR**25** # 4216.

8. J. B. Keller, *The geometrical theory of diffraction*, J. Opt. Soc. Amer. **12** (1962.), 116–130.

9. ———, *Diffraction by a convex cyclinder*, Div. Electromag. Res., Inst. Math. Sci., New York Univ., Res. Report No. EM-94, 1956; Trans. I.R.E. **AP-4** (1956), 312–321. MR**20** #641.

10. J. B. Keller, R. M. Lewis and B. D. Seckler, *Asymptotic solution of some diffraction problems*, Comm. Pure Appl. Math. **9** (1956), 207–265. MR**18**, 43.

11. J. B. Keller, *Surface waves on water of non-uniform depth*, J. Fluid Mech. **4** (1958), 607–614. MR**21** #1070.

12. ———, *Water waves produced by earthquakes*, Proc. Conf. on Tsunami Hydrodynamics, Inst. Geophys. **24** (1961), 154.

13. J. B. Keller and V. C. Mow, *Internal wave propagation in an inhomogeneous fluid of non-uniform depth*, J. Fluid Mech. **38** (1969), 365–374. MR**40** #3788.

14. B. R. Levy and J. B. Keller, *Diffraction by a smooth object*, Comm. Pure Appl. Math. **12** (1959), 159–209. MR**21** #1130.

15. ———, *Diffraction by a spheroid*, Canad. J. Phys. **38** (1960), 128–144. MR**22** #1319.

16. B. R. Levy, *Diffraction by an elliptic cylinder*, J. Math. Mech. **9** (1960), 147–165. MR**22** #7649.

17. R. M. Lewis and J. B. Keller, *Asymptotic methods for partial differential equations: the reduced wave equation and Maxwell's equation*, New York University CIMS Report #EM-194, 1964.

18. R. M. Lewis, *Asymptotic methods for the solution of dispersive hyperbolic equations*, Asymptotic Solutions of Differential Equations and Their Applications (Proc. Sympos., Math. Res. Center, Univ. Wisconsin, Madison, Wis., 1964), Wiley, New York, 1964, pp. 53–107. MR**29** #6187.

19. R. M. Lewis and J. Boersma, *Uniform asymptotic theory of edge diffraction*, J. Mathematical Phys. **10** (1969), 2291–2305.

20. R. M. Lewis, N. Bleistein and D. Ludwig, *Uniform asymptotic theory of creeping waves*, Comm. Pure Appl. Math. **20** (1967), 295–328. MR**35** # 3966.

21. D. Ludwig, *Uniform asymptotic expansions at a caustic*, Comm. Pure Appl. Math. **19** (1966), 215–250. MR**33** # 4446.

22. ———, *Uniform asymptotic expansion of the field scattered by a convex object at high frequency*, Comm. Pure Appl. Math. **20** (1967), 103–138. MR**34** # 3879.

23. ———, *Uniform asymptotic expansions for wave propagation and diffraction problems*, SIAM Rev. **12** (1970), 325–331. MR**42** # 1407.

24. B. D. Sechler and J. B. Keller, *Geometrical theory of diffraction in inhomogeneous media*, J. Acoust. Soc. Amer. **31** (1959), 192–205. MR**20** # 6926.

25. ———, *Asymptotic theory of diffraction in inhomogeneous media*, J. Acoust. Soc. Amer. **31** (1959), 206–216. MR**20** # 6927.

26. G. B. Whitham, *Group velocity and energy propagation for three dimensional waves*, Comm. Pure Appl. Math. **14** (1961), 675–691. MR**24** # B1823.

27. ———, *A general approach to linear and non-linear waves using a Lagrangian*, J. Fluid Mech. **22** (1965), 273–283. MR**31** # 6459.

28. ———, *Dispersive waves and variational principles*, Studies in Applied Mathematics, vol. 7, Prentice-Hall, Englewood Cliffs, N. J., 1971, pp. 181–212.

University of Denver

Lectures in Applied Mathematics
Volume 15, 1974

Studies of Nonlinear Problems. I*

E. Fermi, J. Pasta and S. Ulam

Abstract. A one-dimensional dynamical system of 64 particles with forces between neighbors containing nonlinear terms has been studied on the Los Alamos computer MANIAC I. The nonlinear terms considered are quadratic, cubic, and broken linear types. The results are analyzed into Fourier components and plotted as a function of time.

The results show very little, if any, tendency toward equipartition of energy among the degrees of freedom.

The last few examples were calculated in 1955. After the untimely death of Professor E. Fermi in November, 1954, the calculations were continued in Los Alamos.

This report is intended to be the first one of a series dealing with the behavior of certain nonlinear physical systems where the nonlinearity is introduced as a perturbation to a primarily linear problem. The behavior of the systems is to be studied for times which are long compared to the characteristic periods of the corresponding linear problems.

The problems in question do not seem to admit of analytic solutions in closed form, and heuristic work was performed numerically on a fast electronic computing machine (MANIAC I at Los Alamos).[1] The ergodic

AMS (*MOS*) *subject classifications* (1970). Primary 70–04, 70K99, 82–04, 82A30, 82A77; Secondary 80–04, 80A35.

* This article, reproduced from Los Alamos Report LA1940 (1955), posed the problem and laid some of the groundwork for recent work in nonlinear wave theory, including some of the work reported in this volume. There is no doubt that, were the authors to rewrite the report today, they would make substantive changes. However, we felt that the original article has historical value and because it is not generally available, we decided, with the permission of John Pasta and Stan Ulam, to reproduce it here without change.

[1] We thank Miss Mary Tsingou for efficient coding of the problems and for running the computations on the Los Alamos MANIAC machine.

behavior of such systems was studied with the primary aim of establishing, experimentally, the rate of approach to the equipartition of energy among the various degrees of freedom of the system. Several problems will be considered in order of increasing complexity. This paper is devoted to the first one only.

We imagine a one-dimensional continuum with the ends kept fixed and with forces acting on the elements of this string. In addition to the usual linear term expressing the dependence of the force on the displacement of the element, this force contains higher order terms. For the purposes of numerical work this continuum is replaced by a finite number of points (at most 64 in our actual computation) so that the partial differential equation defining the motion of this string is replaced by a finite number of total differential equations. We have, therefore, a dynamical system of 64 particles with forces acting between neighbors with fixed end points. If x_i denotes the displacement of the ith point from its original position, and α denotes the coefficient of the quadratic term in the force between the neighboring mass points and β that of the cubic term, the equations were either

$$\ddot{x}_i = (x_{i+1} + x_{i-1} - 2x_i) + \alpha[(x_{i+1} - x_i)^2 - (x_i - x_{i-1})^2], \quad i = 1, 2, \ldots, 64. \tag{1}$$

or

$$\ddot{x}_i = (x_{i+1} + x_{i-1} - 2x_i) + \beta[(x_{i+1} - x_i)^3 - (x_i - x_{i-1})^3], \quad i = 1, 2, \ldots, 64. \tag{2}$$

α and β were chosen so that at the maximum displacement the nonlinear term was small, e.g., of the order of one-tenth of the linear term. The corresponding partial differential equation obtained by letting the number of particles become infinite is the usual wave equation plus nonlinear terms of a complicated nature.

Another case studied recently was

$$\ddot{x}_i = \delta_1(x_{i+1} - x_i) - \delta_2(x_i - x_{i-1}) + c, \tag{3}$$

where the parameters δ_1, δ_2, c were not constant but assumed different values depending on whether or not the quantities in parentheses were less than or greater than a certain value fixed in advance. This prescription amounts to assuming the force as a broken linear function of the displacement. This broken linear function imitates to some extent a cubic dependence. We show the graphs representing the force as a function of displacement in three cases.

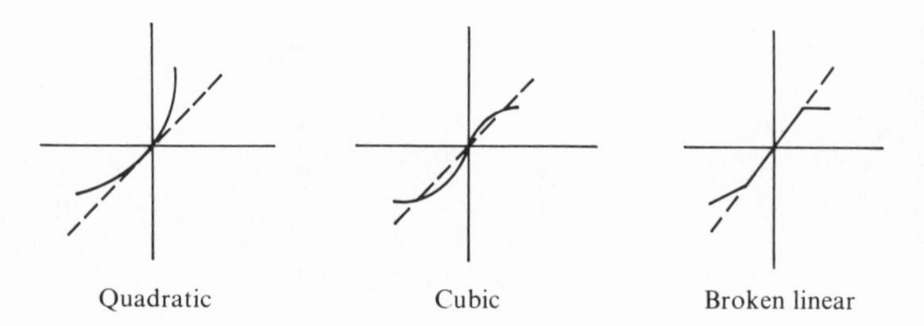

The solution to the corresponding linear problem is a periodic vibration of the string. If the initial position of the string is, say, a single sine wave, the string will oscillate in this mode indefinitely. Starting with the string in a simple configuration, for example in the first mode (or in other problems, starting with a combination of a few low modes), the purpose of our computations was to see how, due to nonlinear forces perturbing the periodic linear solution, the string would assume more and more complicated shapes, and, for t tending to infinity, would get into states where all the Fourier modes acquire increasing importance. In order to see this, the shape of the string, that is to say, x as a function of i and the kinetic energy as a function i, were analyzed periodically in Fourier series. Since the problem can be considered one of dynamics, this analysis amounts to a Lagrangian change of variables: instead of the original $\dot{x}_i$ and x_i, $i = 1, 2, \ldots, 64$, we may introduce a_k and $\dot{a}_k$, $k = 1, 2, \ldots, 64$, where

$$a_k = \sum x_i \sin \frac{ik\pi}{64}. \tag{4}$$

The sum of kinetic and potential energies in the problem with a quadratic force is

$$E_{x_i}^{\text{kin}} + E_{x_i}^{\text{pot}} = \frac{1}{2}\dot{x}_i^2 + \frac{(x_{i+1} - x_i)^2 + (x_i - x_{i-1})^2}{2}, \tag{5a}$$

$$E_{a_k}^{\text{kin}} + E_{a_k}^{\text{pot}} = \frac{1}{2}\dot{a}_k^2 + 2a_k^2 \sin^2 \frac{\pi k}{128}, \tag{5b}$$

if we neglect the contributions to potential energy from the quadratic or higher terms in the force. This amounts in our case to at most a few percent.

The calculation of the motion was performed in the x variables, and every few hundred cycles the quantities referring to the a variables were computed by the above formulas. It should be noted here that the calculation of the motion could be performed directly in a_k and $\dot{a}_k$. The formulas, however, become unwieldy and the computation, even on an electronic computer, would take a long time. The computation in the a_k variables could have been more instructive for the purpose of observing directly the interaction between the a_k's. It is proposed to do a few such calculations in the near future to observe more directly the properties of the equations for $\ddot{a}_k$.

Let us say here that the results of our computations show features which were, from the beginning, surprising to us. Instead of a gradual, continuous flow of energy from the first mode to the higher modes, all of the problems show an entirely different behavior. Starting in one problem with a quadratic force and a pure sine wave as the initial position of the string, we indeed observe initially a gradual increase of energy in the higher modes as predicted (e.g., by Rayleigh in an infinitesimal analysis). Mode 2 starts increasing first, followed by mode 3, and so on. Later on, however, this gradual sharing of energy among successive modes ceases. Instead, it is one or the other mode that predominates. For example, mode 2 decides, as it were, to increase rather rapidly at the cost of all other modes and becomes predominant. At one time, it has more energy than all the others put together! Then mode 3 undertakes this rôle. It is only the first few modes which exchange energy among themselves and they do this in a rather regular fashion. Finally, at a later time mode 1 comes back to within one percent of its initial value so that the system seems to be almost periodic. All our problems have at least this one feature in common. Instead of gradual increase of all the higher modes, the energy is exchanged, essentially, among only a certain few. It is, therefore, very hard to observe the rate of "thermalization" or mixing in our problem, and this was the initial purpose of the calculation.

If one should look at the problem from the point of view of statistical mechanics, the situation could be described as follows: the phase space of a point representing our entire system has a great number of dimensions. Only a very small part of its volume is represented by the regions where only one or a few out of all possible Fourier modes have divided among themselves almost all the available energy. If our system with nonlinear forces acting between the neighboring points should serve as a good example of a transformation of the phase space which is ergodic or metrically transitive, then the trajectory of almost every point should be everywhere dense in the whole phase space. With overwhelming probability this should also be true of the point which at time $t = 0$ represents our initial configuration, and this point should spend most of its time in regions corresponding to the equipartition of energy among various degrees of freedom. As will be seen from the results this seems hardly the case. We have plotted (Figures 1 to 7) the ergodic sojourn times in certain subsets of our phase space. These may show a tendency to approach limits as guaranteed by the ergodic theorem. These limits, however, do not seem to correspond to equipartition even in the time average. Certainly, there seems to be very little, if any, tendency towards equipartition of energy among all degrees of freedom at a given time. In other words, the systems certainly do not show mixing.[2]

The general features of our computation are these: in each problem, the system was started from rest at time $t = 0$. The derivatives in time, of course, were replaced for the purpose of numerical work by difference expressions. The length of time cycle used varied somewhat from problem to problem. What corresponded in the linear problem to a full period of the motion was divided into a large number of time cycles (up to 500) in the computation. Each problem ran through many "would-be periods" of the linear problem, so the number of time cycles in each computation ran to many thousands. That is to say, the number of swings of the string was of the order of several hundred, if by a swing we understand the period of the initial configuration in the corresponding linear problem. The distribution of energy in the Fourier modes was noted after every few hundred of the computation cycles. The accuracy of the numerical work was checked by the constancy of the quantity representing the total energy. In some cases, for checking purposes, the corresponding linear problems were run and these behaved correctly within one percent or so, even after 10,000 or more cycles.

It is not easy to summarize the results of the various special cases. One feature which they have in common is familiar from certain problems in mechanics of systems with a few degrees of freedom. In the compound pendulum problem one has a transformation of energy from one degree of freedom to another and back again, and not a continually increasing sharing of energy between the two. What is perhaps surprising in our problem is that this kind of behavior still appears in systems with, say, 16 or more degrees of freedom.

What is suggested by these special results is that in certain problems which are approximately linear the existence of quasi-states may be conjectured.

In a linear problem the tendency of the system to approach a fixed "state" amounts, mathematically, to convergence of iterates of a transformation in accordance with an algebraic theorem due to Frobenius and Perron. This theorem may be stated roughly in the following way. Let A be a matrix with positive elements. Consider the linear transformation of the n-dimensional space defined by this matrix. One can assert

[2] One should distinguish between metric transitivity or ergodic behavior and the stronger property of mixing.

that if $\bar{x}$ is any vector with all of its components positive, and if A is applied repeatedly to this vector, the directions of the vectors $\bar{x}$, $A(\bar{x}), \ldots, A^i(\bar{x}), \ldots$ will approach that of a fixed vector $\bar{x}_0$ in such a way that $A(\bar{x}_0) = \lambda(\bar{x}_0)$. This eigenvector is unique among all vectors with all their components nonnegative. If we consider a linear problem and apply this theorem, we shall expect the system to approach a steady state described by the invariant vector. Such behavior is in a sense diametrically opposite to an ergodic motion and is due to a very special character, linearity of the transformations of the phase space. The results of our calculation on the nonlinear vibrating string suggest that in the case of transformations which are approximately linear, differing from linear ones by terms which are very simple in the algebraic sense (quadratic or cubic in our case), something analogous to the convergence to eigenstates may obtain.

One could perhaps conjecture a corresponding theorem. Let Q be a transformation of an n-dimensional space which is nonlinear but is still rather simple algebraically (let us say, quadratic in all the coordinates). Consider any vector $\bar{x}$ and the iterates of the transformation Q acting on the vector $\bar{x}$. In general, there will be no question of convergence of these vectors $Q^n(\bar{x})$ to a fixed direction.

But a weaker statement is perhaps true. The directions of the vectors $Q^n(\bar{x})$ sweep out certain cones C_α or solid angles in space in such a fashion that the time averages, i.e., the time spent by $Q^n(\bar{x})$ in C_α, exist for $n \to \infty$. These time averages may depend on the initial $\bar{x}$ but are able to assume only a finite number of different values, given C_α. In other words, the space of all directions divides into a finite number of regions R_i, $i = 1, \ldots, k$, such that for vectors $\bar{x}$ taken from any one of these regions the percentage of time spent by images of $\bar{x}$ under the Q^n are the same in any C_α.

The graphs which follow show the behavior of the energy residing in various modes as a function of time; for example, in Figure 1 the energy content of each of the first five modes is plotted. The abscissa is time measured in computational cycles, δt, although figure captions give δt^2 since this is the term involved directly in the computation of the acceleration of each point. In all problems the mass of each point is assumed to be unity; the amplitude of the displacement of each point is normalized to a maximum of 1. N denotes the number of points and therefore the number of modes present in the calculation. α denotes the coefficient of the quadratic term and β that of the cubic term in the force between neighboring mass points.

We repeat in all our problems we started the calculation from the string at rest at $t = 0$. The ends of the string are kept fixed.

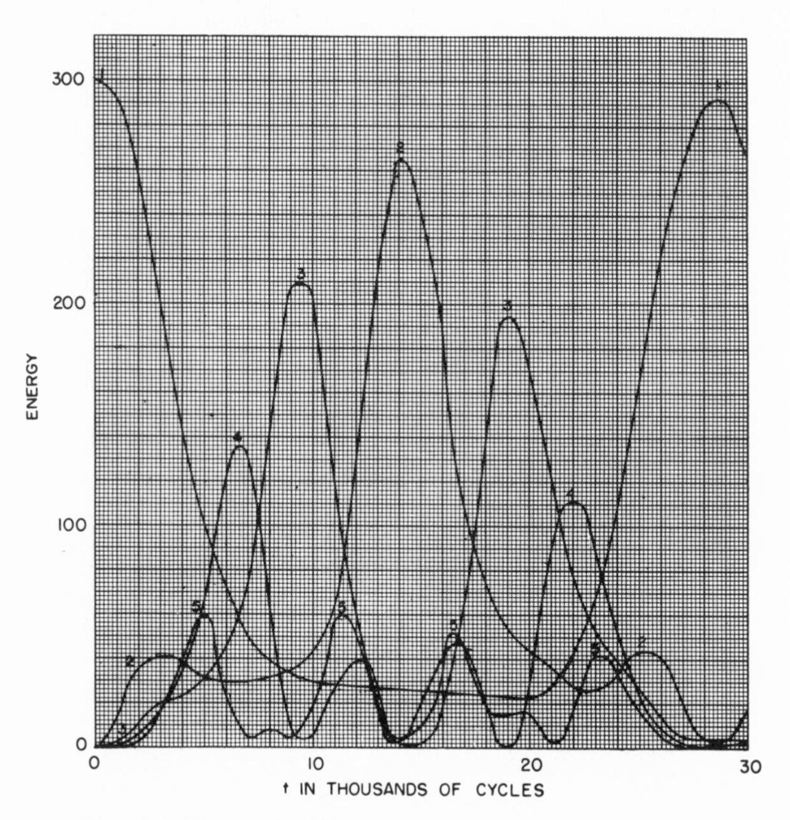

FIGURE 1. The quantity plotted is the energy (kinetic plus potential in each of the first five modes). The units for energy are arbitrary. $N = 32$; $\alpha = 1/4$; $\delta t^2 = 1/8$. The initial form of the string was a single sine wave. The higher modes never exceeded in energy 20 of our units. About 30,000 computation cycles were calculated.

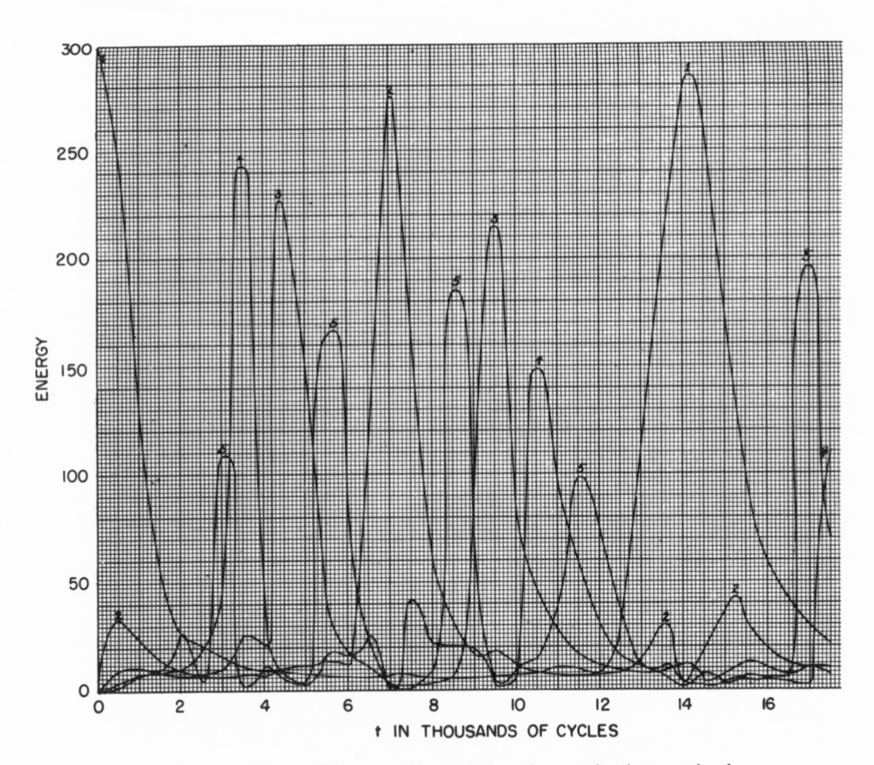

FIGURE 2. Same conditions as Figure 1 but the quadratic term in the force was stronger. $\alpha = 1$. About 14,000 cycles were computed.

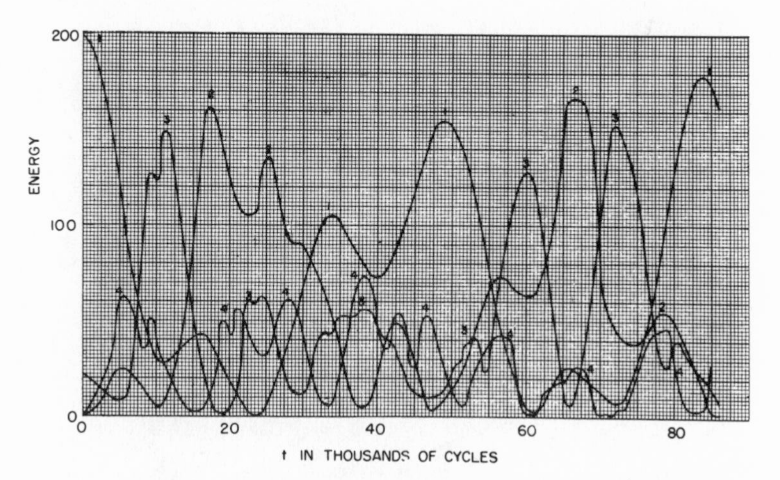

FIGURE 3. Same conditions as in Figure 1, but the initial configuration of the string was a "saw-tooth" triangular-shaped wave. Already at $t = 0$, therefore, energy was present in some modes other than 1. However, modes 5 and higher never exceeded 40 of our units.

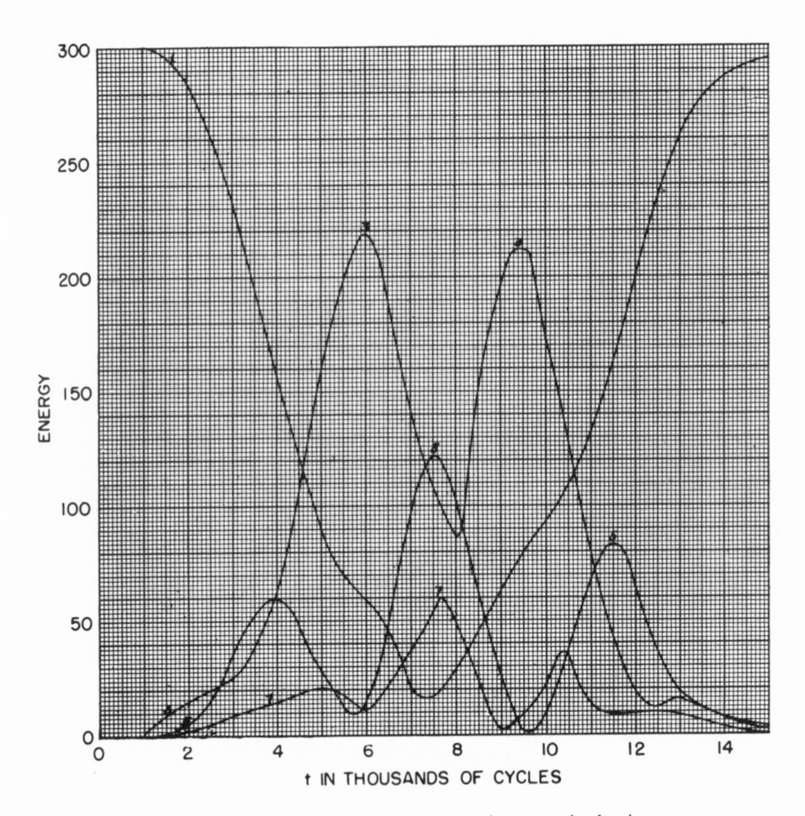

FIGURE 4. The initial configuration assumed was a single sine wave; the force had a cubic term with $\beta = 8$ and $\delta t^2 = 1/8$. Since a cubic force acts symmetrically (in contrast to a quadratic force), the string will forever keep its symmetry and the effective number of particles for the computation $N = 16$. The even modes will have energy 0.

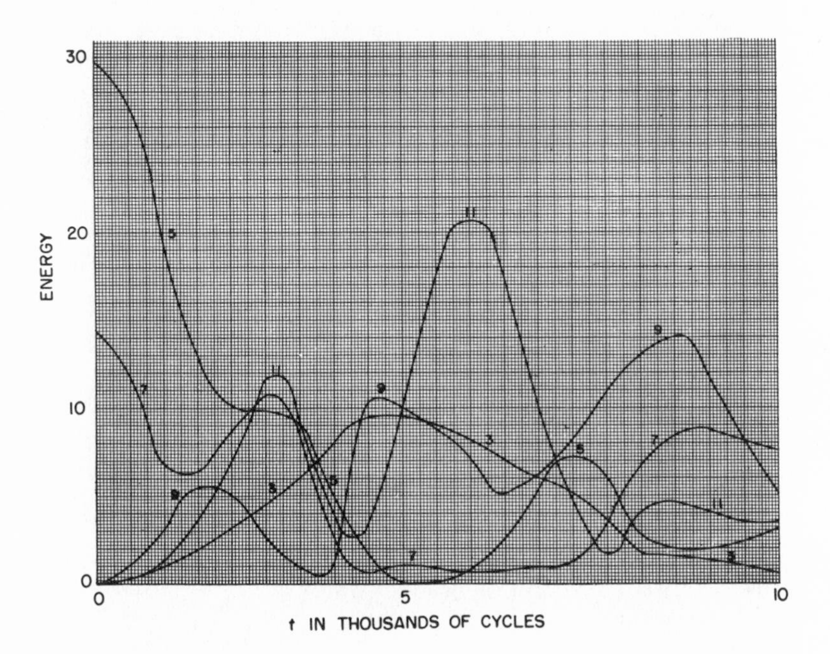

FIGURE 5. $N = 32$; $\delta t^2 = 1/64$; $\beta = 1/16$. The initial configuration was a combination of two modes. The initial energy was chosen to be 2/3 in mode 5 and 1/3 in mode 7.

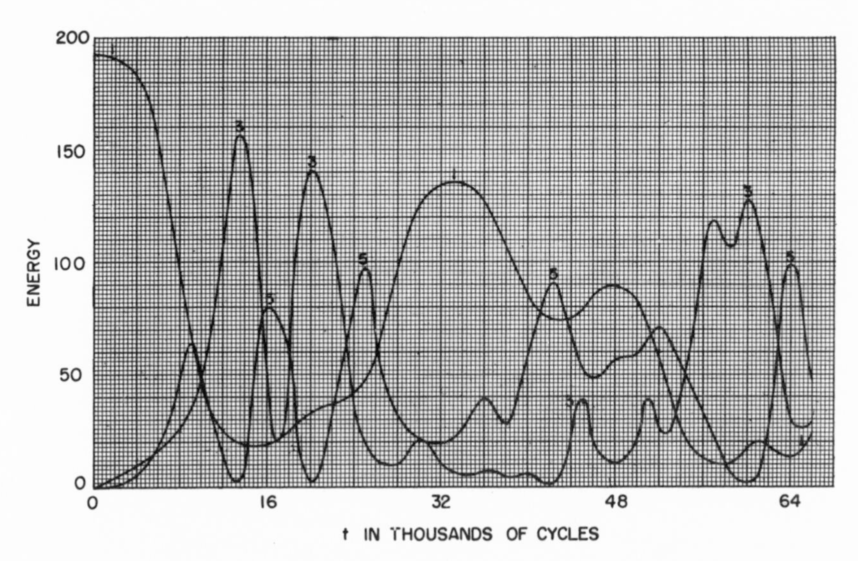

FIGURE 6. $\delta t^2 = 2^{-6}$. The force was taken as a broken linear function of displacement. The amplitude at which the slope changes was taken as $2^{-5} + 2^{-7}$ of the maximum amplitude. After this cut-off value, the force was assumed still linear but the slope increased by 25 percent. The effective $N = 16$.

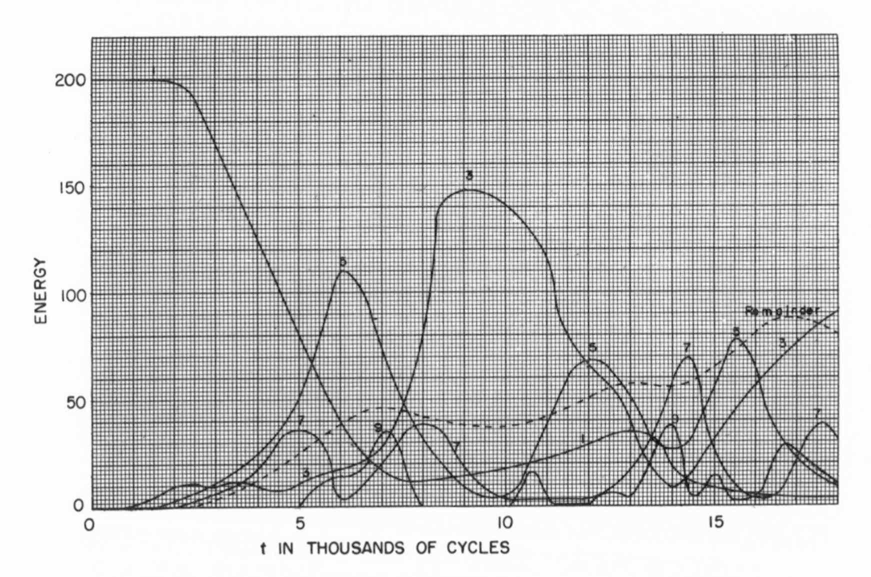

FIGURE 7. $\delta t^2 = 2^{-6}$. Force is again broken linear function with the same cut-off, but the slopes after that increased by 50 percent instead of the 25 percent charge as in Figure 6. The effective $N = 16$.

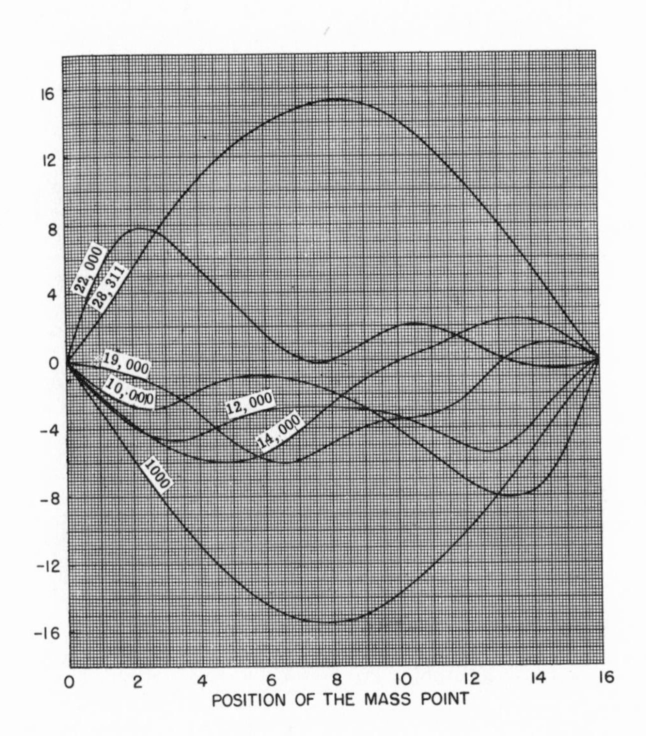

FIGURE 8. This drawing shows not the energy but the actual *shapes*, i.e., the displacement of the string at various times (in cycles) indicated on each curve. The problem is that of Figure 1.

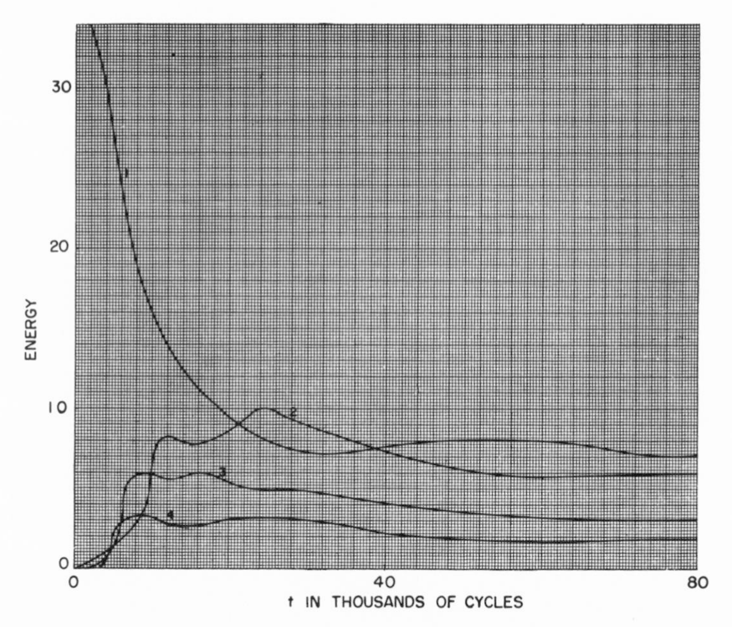

FIGURE 9. This graph refers to the problem of Figure 6. The curves, numbered 1, 2, 3, 4, show the time averages of the kinetic energy contained in the first 4 modes as a function of time. In other words, the quantity is $(1/\nu)\sum_{i=1}^{\nu} T^{i}_{a_k}$. ν is the cycle number, $k = 1, 3, 5, 7$.

Lectures in Applied Mathematics
Volume 15, 1974

Envelope Equations

Alan C. Newell

Abstract. A unified theory is presented for the derivation of equations which describe the slow temporal and spatial modulation of the envelopes of almost monochromatic wave trains. A general equation is found which in various limits reduces to the envelope equations describing (a) neutral, dispersive, weakly nonlinear wave trains, (b) the post bifurcation regime of problems where the principle of exchange of stabilities is valid, (c) the post bifurcation regime when instability sets in as growing oscillations, (d) the post bifurcation regime in essentially dissipation free systems. Some properties of these equations are discussed. Of particular interest is the existence of a general criterion for the stability of a monochromatic wave train analogous to the Benjamin–Feir criterion for instability of water waves.

1. Introduction and derivation of the envelope equation. Physical phenomena are often described by means of nonlinear partial differential equations for which the general initial value problem is yet unsolved. As a result our understanding of such phenomena is rather limited, but considerable progress has been made in cases where a basic solution with some determined periodic structure can be found. These special solutions have associated with them certain parameters which can be identified with the usual concepts of amplitude, wave number and frequency. It is of considerable interest to describe the evolution of these parameters themselves, assuming that everywhere in space and time the solution is given locally by the predetermined basic structure. We can usually achieve this goal when the generic system has certain properties like small ampli-

AMS (MOS) subject classifications (1970). Primary 34E15, 35F20, 35F25, 76B15.

tudes or a large separation between the relevant scales (both space and time) which appear in the problem.

Consider the nonlinear partial differential equation

$$(1.1) \qquad L(\partial/\partial t, \partial/\partial x, R)w = MwNw, \qquad 0 \leqq t, -\infty < x < \infty,$$

where L, M and N are differential operators and R is a given fixed parameter. The dependence of the operators and in particular the operator L on spatial dimensions which are of finite extent and in which direction the spectrum is quantized is not crucial to the present discussion. A linear stability analysis on the trivial solution of (1.1) may readily be carried out by neglecting the R.H.S. of (1.1) and looking for solutions of the form

$$(1.2) \qquad w \propto e^{ikx-\omega t}, \qquad \omega = \omega_r + i\omega_i.$$

Marginal stability occurs when $\omega_r = 0$, and this leads to certain compatibility relations given by the equation

$$(1.3) \qquad L(-i\omega_i(k), ik, R) = 0$$

which yields the neutral stability curve R vs. k and the companion dispersion relation $\omega_i = \omega_i(k)$. In many physical problems, the neutral stability curve has a minimum R_c at a finite value $k(k_c)$, reflecting the choice of some optimal scale at which the release of potential energy overcomes the inhibiting effects of viscosity, stratification, conductivity, etc. We will assume that the lowest critical R number occurs for $\omega_i \neq 0$ although the results in which the principle of exchange of stabilities holds can be readily deduced from the final envelope equation.

Suppose now we want to describe the motion when

$$(1.4) \qquad R = R_c(1 + \varepsilon^2\bar{\chi}).$$

The fact that the neutral stability curve is parabolic in the neighborhood of (k_c, R_c) means that an $O(\varepsilon)$ band of wave numbers $k_c \pm O(\varepsilon)$ can draw on the source of potential which derives the instability. The wave packet concept is simply incorporated into the description of the post bifurcation stage by writing

$$(1.5) \qquad w \propto W(X, T_1, T_2)\, e^{ik_cx - i\omega_i(k_c)t} + (*), \qquad X = \varepsilon x,$$

where (*) represents the complex conjugate and W is the amplitude function. We next ask: What are the relevant scales? Since we expect wave packets to move with the group velocity until dispersive effects predominate, it is relevant to introduce the time scale $T_1 = \varepsilon t$. Since the initial growth rate of linear disturbances is $O((R - R_c)/R_c)$ we introduce the time scale $T_2 = \varepsilon^2 t$; this is also the scale on which dispersive effects become important. One expects the wave amplitude to grow until inhibited by nonlinear effects. Since nonlinear effects of a growing wave with the structure (1.5) can only manifest themselves through a cubic interaction, we expect an amplitude balance can be achieved when $w = O(\varepsilon)$. Accordingly, we set

$$w = \varepsilon(w_0 + \varepsilon w_1 + \varepsilon^2 w_2 + \cdots) \tag{1.6}$$

where w_0 is given by (1.5). We solve (1.1) recursively and choose at each stage the evolution of the amplitude (or envelope) function $W(X, T_1, T_2)$ in order to suppress nonuniformities which would otherwise appear in (1.6).

If we restrict ourselves to the case of a discrete wave, the amplitude W is only a function of T_2 and its evolution is given by the Stuart–Watson ([**1**], [**2**]) equation

$$dW/dT_2 = \chi W - \beta W^2 W^*, \qquad \beta = \beta_r + i\beta_i. \tag{1.7}$$

The effects of spatial modulation lead to additional terms, but these terms only arise from the operator L. Introduce the transformations

$$\begin{aligned} \frac{\partial}{\partial t} &\to \frac{\partial}{\partial t} + \varepsilon\frac{\partial}{\partial T_1} + \varepsilon^2\frac{\partial}{\partial T_2}, \\ \frac{\partial}{\partial x} &\to \frac{\partial}{\partial x} + \varepsilon\frac{\partial}{\partial X}, \\ R &\to R_c(1 + \varepsilon^2\bar{\chi}), \end{aligned} \tag{1.8}$$

into the differential operator L and expand by a Taylor series. We obtain

$$\begin{aligned} L \to L\left(\frac{\partial}{\partial t}, \frac{\partial}{\partial x}, R_c\right) &+ \varepsilon\left(L_1\frac{\partial}{\partial T_1} + L_2\frac{\partial}{\partial X}\right) \\ &+ \varepsilon^2\left(\frac{1}{2}L_{11}\frac{\partial^2}{\partial T_1^2} + L_1\frac{\partial}{\partial T_2} + L_{12}\frac{\partial^2}{\partial T_1 \partial X}\right. \\ &\left.\quad + \frac{1}{2}L_{22}\frac{\partial^2}{\partial X^2} + L_2 R_c\bar{\chi}\right) + \cdots \end{aligned} \tag{1.9}$$

where the subscripts (1, 2, 3) denote partial differentiation with respect to the arguments $\partial/\partial t$, $\partial/\partial x$ and R respectively. Given a specific L, one can compute the partial derivatives exactly. A much clearer insight into the nature of the evolution equation is gained by recognizing that we can find the relation between the values of the partial derivatives of L at their critical values $\partial/\partial t = -i\omega_i(k_c)$, $\partial/\partial x = ik_c$, $R = R_c$, by differentiating the complex dispersion relation (1.3) with respect to k.

In order to suppress nonuniformities in w_1, we find

$$L_1\left(\frac{\partial}{\partial T_1} + \frac{d\omega_i}{dk}\frac{\partial}{\partial X}\right)W = 0. \tag{1.10}$$

Thus if L_1 is nonzero at critical $W = W(\bar{X}, T_2), \bar{X} = X - (d\omega_i/dk)T_1$. In order to suppress nonuniformities from appearing in w_2, we have

$$\begin{aligned} &L_1\left(\frac{\partial W}{\partial T_2} - \frac{i}{2}\frac{d^2\omega_i}{dk^2}\frac{\partial^2 W}{\partial \bar{X}^2}\right) \\ &\qquad + L_3\left(\frac{1}{2}\frac{d^2R}{dk^2}\frac{\partial^2 W}{\partial \bar{X}^2} + R_c\bar{\chi}W\right) = -\bar{\beta}W^2W^*, \end{aligned} \tag{1.11}$$

where the nonlinear terms $\bar{\beta}W^2W^*$ are the same as calculated for the Stuart–Watson equation. All other coefficients are functions of k and estimated at k_c. Equation (1.11) was originally derived by Newell and Whitehead ([**3**], [**4**]) and has been used by Hocking, Stewartson and Stuart ([**5**], [**6**]) and Haberman [**7**] in looking at parallel flow stability problems.

If L_1 is zero, then (1.10) is trivially satisfied and equation (1.11) is modified to be

$$\begin{aligned} &\frac{1}{2}L_{11}\left(\frac{\partial^2}{\partial T_1^2} - \left(\frac{d\omega_i}{dk}\right)^2\frac{\partial^2}{\partial X^2}\right)W + L_{12}\frac{\partial}{\partial X}\left(\frac{\partial}{\partial T_1} - \frac{d\omega_i}{dk}\frac{\partial}{\partial X}\right)W \\ &\qquad + L_3\left(\frac{1}{2}\frac{d^2R}{dk^2}\frac{\partial^2}{\partial X^2} + R_c\bar{\chi}\right)W = -\bar{\beta}W^2W^* \end{aligned} \tag{1.12}$$

which has the structure of a nonlinear Klein–Gordon equation. Such a case arises when $\omega_i(k_c)$ is a double root of the linear operator, and examples occur in the problems of the buckling of spherical and cylindrical shells under radial and axial loads [**8**] and in the inviscid treatment of the baroclinic instability [**9**].

We have provided a unified theory for the temporal and spatial modulation of almost monochromatic wave trains without appeal to specific operators L. However, a word of caution. When are such descriptions relevant? *The underlying assumption is that the basic structure is given by the eigenfunctions of the linear problem.* One expects, therefore, that the theory is applicable in problems where small amplitude theory is relevant.

2. Properties of the envelope equation. We rewrite (1.11) as

$$\begin{aligned} &\frac{\partial W}{\partial t} - \gamma\frac{\partial^2 W}{\partial x^2} = \chi W - \beta W^2 W^*, \qquad \gamma = \gamma_r + i\gamma_i, \\ &\qquad \beta = \beta_r + i\beta_i, \qquad \chi \text{ real}, \end{aligned} \tag{2.1}$$

and consider some limiting cases. If $\gamma_r = \beta_r = \chi = 0$, then (2.1) describes the evolution of the envelope of a weakly nonlinear dispersive wave which is neutral in character.

$$\partial W/\partial t - i\gamma_i \partial^2 W/\partial x^2 = -i\beta_i W^2 W^*. \tag{2.2}$$

Equation (2.2) has been derived in different contexts: (a) as the spanwise modulation of the electric field in a medium where the refractive index depends on the amplitude ([**10**], [**11**]), (b) as the envelope equation ([**12**], [**13**]). It has the interesting property that x independent solutions are unstable when $\beta_i\gamma_i < 0$ (analogous to the Benjamin–Feir [**14**] criterion for the instability of gravity water waves). In this case, (2.2) has been solved exactly by Zakharov and Shabat [**15**] using the inverse scattering method, and the final state is a sequence of individual or paired permanent waves superimposed on a background field of decaying (algebraically in time) self-similar structure [**16**]. The velocities are determined by the eigenvalues of the appropriate scattering problem. If $\beta_i\gamma_i > 0$, there are no bound states (permanent waves) and the final state is just the similarity solution [**17**].

Another limit ([**4**], [**18**]) of (2.1) arises when the principle of exchange of stabilities is valid, i.e. $\gamma_i = \beta_i = 0$,

$$\partial W/\partial t - \gamma_r \partial^2 W/\partial x^2 = \chi w - \beta_r W^2 W^*. \tag{2.3}$$

Numerical experiments on (2.2) (with periodic boundary conditions in x) suggest that the amplitude first evolves into a square wave form with the solutions $W = \pm(\chi/\beta_r)^{1/2}$ pieced together by boundary layer solutions $W = (\chi/\beta_r)^{1/2}\tanh(\chi/2\gamma_r)^{1/2}(x - x_0)$. The boundary layer solutions are unstable unless W is real which is the case when $k_c = 0$. Even in this latter case, however, the kinks move slowly and eventually annihilate each other and the final state of the system seems to be either $W = +(\chi/\beta_r)^{1/2}$ or $W = -(\chi/\beta_r)^{1/2}$ everywhere.

This is consistent with the results of the statistical initial value problem (Newell, Lange and Aucoin [**19**]) in which the evolution of an initially small, random disturbance field is examined.

The full equation (2.1) has similar properties. First, it can be verified that the solution

$$W = (\chi/\beta_r)^{1/2}\exp(-i\beta_i(\chi/\beta_r)t) \tag{2.4}$$

is unstable to x dependent disturbances when

$$\beta_r\gamma_r + \beta_i\gamma_i < 0. \tag{2.5}$$

This is the direct analogy with the Benjamin–Feir criterion.

Similarly, (2.1) can be examined from the stochastic viewpoint. Lange and Newell [**20**] have obtained the following results.

When $\beta_r > 0$ and $\beta_r\gamma_r + \beta_i\gamma_i \geqq 0$, the mean field grows from an initial amplitude of μ to a finite quasi-steady value of order unity (given by (2.3)). The second order cumulants grow from an initial size of μ^2 to a size of order $v(\mu)$ ($v(\mu) = (\log(1/\mu))^{-1/2}$) after a time $1/v^2$ and then decay (if $\beta_r\gamma_r + \beta_i\gamma_i > 0$) with time to zero. Likewise the nth order cumulant grows to a size $v^{n-1}(\mu)$ at time $1/v^2$ and then decays to zero. Thus the final state is a perfectly ordered field corresponding to a fixed amplitude, monochromatic wave with an amplitude dependent frequency modulation.

When $\beta_r > 0$ and $\beta_r\gamma_r + \beta_i\gamma_i < 0$, the second and higher order cumulants grow exponentially in the long time limit. This suggests that even though the fundamental wave number initially outgrows its sidebands, it eventually succumbs to their influence and thus (2.4) is the direct analogy to the Benjamin–Feir criterion.

When $\beta_r < 0$, then the solution becomes unbounded in finite time. These results are consistent with those found by Stuart, Stewartson and Hocking and Haberman who examined the problem when the initial conditions on W belong to the class of square integrable functions. No analogy to the Benjamin–Feir criterion seems to exist in this case.

We might next ask: To what state or states does (2.1) evolve when $\beta_r\gamma_r + \beta_i\gamma_i < 0$? We will exhibit two classes of permanent wave solutions of the form $W = re^{i\theta}$. The first solution is (we take χ complex)

$$
\begin{aligned}
r &= a \operatorname{sech} b(x - Vt), \\
\theta_x &= \alpha b \tanh b(x - Vt) + K, \qquad (2.6) \\
\theta_t &= -\alpha b V \tanh b(x - Vt),
\end{aligned}
$$

where

$$
\begin{aligned}
b^2|\gamma|^2(2 - \alpha^2) &= -(\beta_r\gamma_r + \beta_i\gamma_i)a^2, \\
3\alpha b^2|\gamma|^2 &= -(\beta_r\gamma_i - \beta_i\gamma_r)a^2, \\
b^2|\gamma|^2(1 - \alpha^2 - K^2/b^2) &= -(\chi_r\gamma_r + \chi_i\gamma_i), \\
2\alpha b^2|\gamma|^2 &= \chi_r\gamma_i - \chi_i\gamma_r.
\end{aligned}
$$

We observe in particular that if $\gamma_r \neq 0$, then $V = K = 0$ suggesting that the only permanent wave form moves with the group velocity of the fundamental wave k_c. If $\gamma_r = 0$, then $K = V/2\gamma_i$. If $\beta_r = \gamma_r = \chi_r = 0$, then $\alpha = 0$ and $a = ((-2\gamma_i/\beta_i)b)^{1/2}$ and the permanent waves agree with those found from the exact solution of (2.2).

The second solution is given by

$$\begin{aligned} r &= a \tanh b(x - Vt), \\ \theta_x &= \alpha b \tanh b(x - Vt) + K, \\ \theta_t &= -\alpha b V \tanh b(x - Vt). \end{aligned} \tag{2.7}$$

Again, $V = K = 0$ and

$$\begin{aligned} b^2|\gamma|^2(2 - \alpha^2) &= (\beta_r\gamma_r + \beta_i\gamma_i)a^2, \\ 3\alpha b^2|\gamma|^2 &= (\beta_r\gamma_i - \beta_i\gamma_r)a^2, \\ 2b^2|\gamma|^2 &= \chi_r\gamma_r + \chi_i\gamma_i, \\ 3\alpha b^2|\gamma|^2 &= \chi_r\gamma_i - \chi_i\gamma_r. \end{aligned}$$

Note that when $\chi_r = \beta_r = \gamma_r = 0$ solutions only exist when $\beta_i\gamma_i > 0$. Also when $\beta_i = \gamma_i = 0$, solutions exist for $\beta_r\gamma_r > 0$ and have been discussed [**4**].

REFERENCES

1. J. T. Stuart, J. Fluid Mech. **9** (1960), 353–370. MR**23** #B1272.

2. J. Watson, J. Fluid Mech. **9** (1960a), 371–389. MR**23** #B1273.

3. A. C. Newell and J. A. Whitehead, Proceedings IUTAM Symposium on instability of continuous systems (Harrenalb, 1969), Springer-Verlag, New York, 1971, pp. 284–289.

4. ———, J. Fluid Mech. **38** (1969), 279–304.

5. J. Stewartson and J. T. Stuart, J. Fluid. Mech. **48** (1971), 529–545.

6. L. M. Hocking, K. Stewartson and J. T. Stuart, J. Fluid Mech. **51** (1972), 705–735.

7. R. Haberman, J. Fluid Mech. (to appear).

8. C. G. Lange and A. C. Newell, SIAM. J. Appl. Math. **21** (1971), 605–629.

9. J. Pedlosky, J. Atmos. Sci. **29** (1972), 680–686.

10. V. I. Bespalov, A. G. Litvak and V. I. Talanov, Second All-Union Symposium on Nonlinear Optics (1966), Collection of Papers, Nauka, Moscow, 1968.

11. P. L. Kelley, Phys. Rev. Lett. **15** (1965), 1005.

12. D. J. Benney and A. C. Newell, J. Mathematical Phys. **46** (1967), 133–139.

13. J. Wanatabe, J. Phys. Soc. Japan **27** (1969), 1341–1350.

14. T. B. Benjamin and J. E. Feir, J. Fluid Mech. **27** (1966), 417–430.

15. V. E. Zakharov and A. B. Shabat, JETP **34** (1972), 62–69.

16. M. J. Ablowitz and A. C. Newell, J. Mathematical Phys. **14** (1973), 1277–1284.

17. M. J. Ablowitz, D. J. Kaup, A. C. Newell and H. Segur, Phys. Rev. Lett. **31** (1973), 125–127.

18. L. A. Segel, J. Fluid Mech. **38** (1969), 203–224.

19. A. C. Newell, C. G. Lange and P. J. Aucoin, J. Fluid Mech. **40** (1970), 513–542.

20. C. G. Lange and A. C. Newell, SIAM J. Appl. Math. (to appear).

CLARKSON COLLEGE OF TECHNOLOGY

Lectures in Applied Mathematics
Volume 15, 1974

An Alternative Method to Solve the Korteweg–de Vries Equation?

Harvey Segur

Very few nonlinear partial differential equations can be solved exactly for arbitrarily prescribed initial data. Of the equations that can be solved exactly, the form

$$\frac{\partial u}{\partial t} + au\frac{\partial u}{\partial x} + b\frac{\partial^n u}{\partial x^n} = 0 \tag{1}$$

seems to be particularly important. Examples of (1) include the nonlinear wave equation (here subscripts denote differentiation)

$$u_t + uu_x = 0, \tag{2}$$

an equivalent equation

$$V_t + VV_x + bV_x = 0, \tag{3}$$

which can be reduced to (2) by the transformation $v(x, t) + b = u(x, t)$, Burgers' equation

$$u_t + uu_x = \alpha u_{xx}, \qquad \alpha > 0, \tag{4}$$

and the Korteweg–de Vries equation

$$u_t + 6uu_x + u_{xxx} = 0. \tag{5}$$

That all of these equations, and almost no others, can be solved exactly suggests that a single algorithm might solve every equation of the form (1). (*Note added in proof.* Many others have since been found.)

AMS (MOS) subject classifications (1970). Primary 35–02, 35C10, 35F25, 45E99.

Strangely, the methods of solution of (2), (4) and (5), as well as the solutions themselves, show no apparent interrelationship. (2) is solved by characteristics; neither (4) nor (5) is. The inverse-scattering method of solving (5) is linked to the existence of an infinite set of conservation laws. (4) has only one known conservation law, the equation itself.

The purpose of this paper is to explore the possibility of a single algorithm to solve equations of the form (1). In particular, we seek to obtain the exact solution of (5) by means of the procedure used to solve (4). This method reproduces parts of the known exact solution of (5) on the infinite domain, $-\infty < x < \infty$. It appears that it might also generate the solution on the finite domain, $a < x < b$.

The inverse-scattering method of solution of (5) was discovered by Gardner, Greene, Kruskal and Miura [**1967**]. Let us review briefly the principal ingredients of that method. Details may be found in Gardner et al. [**1967**] or Segur [**1973**]. If the initial data for (5) satisfy

$$\int_{-\infty}^{\infty} (1 + |x|)|u(x, 0)|\, dx < \infty,$$

then $u(x, 0)$ generates a "reflection coefficient":

$$B(2x, 0) = \frac{1}{2\pi}\int_{-\infty}^{\infty} b(k)\, e^{2ikx}\, dk + \sum_{n=1}^{N} c_n^2\, e^{-2\kappa_n x} = B_c(2x) + B_D(2x). \tag{6}$$

For $t > 0$, $B(2x, t)$ evolves such that

$$L(B) = B_t + B_{xxx} = 0. \tag{7}$$

Thus, $B(2x, t)$ satisfies the linearized version of (5). Then $K(x, y, t)$ is determined by

$$K(x, y) + B(x + y) + \int_{x}^{\infty} K(x, z)B(z + y)\, dz = 0, \qquad y > x, \tag{8}$$

and the solution of (5) is given by

$$u(x, t) = 2\frac{d}{dx}K(x, x; t). \tag{9}$$

If the initial spectrum is purely discrete, $b(k) \equiv 0$, the integral equation is degenerate, and may be solved in closed form:

$$K_D(x, x; t) = \frac{d}{dx}\ln\det\left(\delta_{mn} + \frac{\tilde{c}_m\tilde{c}_n \exp(-(\kappa_m + \kappa_n)x)}{\kappa_n + \kappa_m}\right), \tag{10}$$

where $\tilde{c}_n = c_n \exp(4\kappa_n^3 t)$, $n = 1, 2, \ldots, N$. Here, N is the number of discrete eigenvalues corresponding to the initial data, and the number of solitons that eventually emerge. We define

$$f_n = \frac{\tilde{c}_n^2 \exp(-2\kappa_n x)}{2\kappa_n} = \frac{c_n^2 \exp(8\kappa_n^3 t - 2\kappa_n x)}{2\kappa_n} \tag{11}$$

and note explicitly, for $N = 1$,

$$K_D = \frac{d}{dx} \ln(1 + f_1); \tag{12a}$$

for $N = 2$,

$$K_D = \frac{d}{dx} \ln\left(1 + f_1 + f_2 + \left(\frac{\kappa_1 - \kappa_2}{\kappa_1 + \kappa_2}\right)^2 f_1 f_2\right). \tag{12b}$$

If the initial spectrum is purely continuous, then $N = 0$. In this case, one can show that the integral term in (8) is small in Hilbert space norm; i.e.,

$$\left\| \int_x^\infty B(y + z)\phi(z)\, dz \right\| < \|\phi\|$$

for any square-integrable function $\phi(y)$. It follows that (8) has a convergent Neumann series solution,

$$\begin{aligned} K_c(x, x) = {} & -B(2x) + \int_x^\infty B(x + z)B(z + x)\, dx \\ & - \iint_x^\infty B(x + z_1)B(z_1 + z_2)B(z_2 + x)\, dz_1\, dz_2 + \cdots . \end{aligned} \tag{13}$$

The general solution obtained by this method is a nonlinear combination of these two special solutions. No new ingredients are added.

Consider next Burgers' equation (4) and the corresponding linear equation

$$\tilde{L}(\theta) = \theta_t - \alpha\theta_{xx} = 0. \tag{14}$$

Hopf [**1950**] and Cole [**1951**] independently solved (4). Cole observed that $\theta(x, t)$ is invariant under the transformation

$$x \to ax, \qquad t \to a^2 t, \tag{15}$$

whereas $u(x, t)$ is not. For u small enough, however, the evolution of u and θ should be similar, since their respective equations are approximately equal. Hence, we rewrite (4) in a form that is invariant under (15). Define $\psi(x, t)$ by

$$u(x, t) = -2\alpha(\ln \psi)_x . \tag{16}$$

Then (4) becomes

(4a) $$\psi(\tilde{L}\psi)_x - \psi_x(\tilde{L}\psi) = 0.$$

$\psi(x, t)$ is invariant under (15). More importantly, if ψ is any solution of (14), then ψ satisfies (4a) and u satisfies (4). Thus, the invariance under (15) leads to the exact solution.

Let us apply the same procedure to the Korteweg–de Vries equation (5) and its linearized form (7). Solutions of (7) are invariant under

(17) $$x \to ax, \qquad t \to a^3 t,$$

whereas solutions of (5) are not. To preserve this invariance, define $F(x, t)$ by

(18) $$u(x, t) = (\ln F)_{xx}.$$

$F(x, t)$ satisfies the equation

(19) $$F(LF)_x - F_x(LF) + 3(F_{xx}^2 - F_x F_{xxx}) = 0,$$

and is invariant under (17). Equation (19) was derived independently by Whitham [**1971**] and by Hirota [**1971**], and was used by each to derive the N-soliton solution of (5). Even though (19) is nonlinear, we note that if $F_1(x, t)$ solves (19), so does $cF_1(x, t)$ for any constant c. Moreover, the role of the linear operator (LF) is exposed, as is the role of exponentials (in x). We shall construct solutions of (19), and thereby solve (5), from solutions of (9).

Consider first the solitons. Special solutions of (7) include the constant solution, $F = 1$, and $f_n(x, t) = a_n \exp(8\kappa_n^3 t - 2\kappa_n x)$. Thus if

(20) $$F(x, t) = 1 + f_1(x, t),$$

then $LF = 0$. But F also solves (19), and actually represents one soliton, with speed $4\kappa_1^2$.

The remarkable thing about solitons is that they can be combined. For two solitons, Whitham [**1971**] sought a solution of (19) as a "perturbation expansion",

(21) $$F(x, t) = 1 + \varepsilon(f_1 + f_2) + \varepsilon^2 h_2(x, t) + \varepsilon^3 h_3(x, t) + \cdots.$$

Substituting into (19), and collecting powers of ε, one obtains a hierarchy of equations:

$$L(f_1 + f_2) = 0,$$

$$L(h_2)_x = 48\kappa_1\kappa_2(\kappa_1 - \kappa_2)^2 f_1 f_2, \qquad (Lh_3)_x = \cdots.$$

These can be solved easily:

$$h_2(x, t) = \left(\frac{\kappa_1 - \kappa_2}{\kappa_1 + \kappa_2}\right)^2 f_1 f_2, \qquad h_3 = 0 = h_4 = h_5 = \cdots.$$

Thus, the expansion (21) terminates at the third term, and reproduces the 2-soliton solution (12b). In general, an expansion that begins with a sum of N exponential solutions of (7),

$$F(x, t) = 1 + \varepsilon(f_1 + f_2 + \cdots + f_N) + \varepsilon^2 h_2(x, t) + O(\varepsilon^3),$$

terminates after $(N + 1)$ terms and gives the N-soliton solution of (5). Thus, the exact solution of (5) associated with the discrete spectrum is found from (19) by a regular perturbation.

The difficult part of the exact solution of (5) is associated with the continuous spectrum, which we consider next. $K(x, y)$ is the solution of an integral equation, (8), of the form

$$(I + \dot{B})K(x, y) = -B(x + y), \tag{8a}$$

where $\dot{B}\phi(y) = \int_x^\infty B(y + z)\phi(z)\, dz$. Can K be represented easily as

$$K(x, x; t) = (\partial/\partial x)(\ln F),$$

where F solves (19)? We proceed formally. The solution of (8a) is

$$K(x, y) = -(I + \dot{B})^{-1}B(x + y).$$

Because of the x-dependence of $(\dot{B})$,

$$\frac{\partial}{\partial x}(\dot{B})\phi \equiv \frac{\partial}{\partial x}\left(\int_x^\infty B(z + y)\,\phi(z)\, dz\right) = -B(x + y)\phi(x).$$

Hence, a formal solution of (8a) is

$$K(x, x; t) = \frac{\partial}{\partial x}(\ln (I + \dot{B})). \tag{22}$$

It remains to interpret (22). Recalling the expansion of $\ln(1 + a)$, we define

$$\begin{aligned}\ln (I + \dot{B})(x) &= \dot{B} - \tfrac{1}{2}(\dot{B})^2 + \tfrac{1}{3}(\dot{B})^3 - \tfrac{1}{4}(\dot{B})^4 + \cdots \\ &= \int_x^\infty B(2z)\, dz - \frac{1}{2}\iint_x^\infty B(z_1 + z_2)B(z_2 + z_1)\, dz_1\, dz_2 \\ &\quad + \frac{1}{3}\iiint_x^\infty B(z_1 + z_2)B(z_2 + z_3) \\ &\qquad B(z_3 + z_1)\, dz_1\, dz_2\, dz_3 - \cdots,\end{aligned} \tag{23}$$

whenever these integrals exist.

Equation (23) defines ($\ln F$) and yields the solution of (5) when $B(x)$ satisfies (7). First, if no solitons are present ($B = B_c$), then

$$K(x, x; t) = \frac{\partial}{\partial x}(\ln(I + \dot{B}))$$

$$= -B(2x) + \int_x^\infty B(x + z_1)B(z_1 + x)\,dz_1$$

$$- \iint_x^\infty B(x + z_2)B(z_2 + z_3)B(z_3 + x)\,dz_2\,dz_3 + \cdots,$$

which is the Neumann series (13). In this case, all the integrals exist, and the series converges. Second, if no continuous spectrum is present ($B = B_D$), then the integrals can be computed explicitly, and formulae like (12) are reproduced. Finally, the formal derivation of (23) remains the same in the general case, when both spectra exist.

The solution of (19) is

$$F = (I + \dot{B}) \equiv e^{\ln(I + \dot{B})}$$

$$(24)\qquad \begin{aligned} F(x, t) = 1 &+ \int_x^\infty B(2z)\,dz \\ &+ \frac{1}{2!}\left[\int_x^\infty B\,dz\right]^2 - \frac{1}{2}\iint_x^\infty B^2\,dz_1\,dz_2 \\ &+ \frac{1}{3!}\left[\int_x^\infty B\,dz\right]^3 - \frac{1}{2}\int_x^\infty B\,dz \iint_x^\infty B^2\,dz_1\,dz_2 \\ &+ \frac{1}{3}\iiint_x^\infty B^3\,dz_1\,dz_2\,dz_3 + \cdots, \end{aligned}$$

where $B(x; t)$ satisfies (7). Again, we note that for a purely discrete spectrum these integrals can be computed explicitly and that F reproduces the exact solution found above. Thus, (24) represents the formal solution of (19), as sought.

The question remains of whether these integrals converge. The expansion in (24) can be identified with the Fredholm expansion of the resolvent kernel of the integral equation (8). The Fredholm theory requires that ($\dot{B}$) be compact, i.e.,

$$(25)\qquad \iint_x^\infty B(z + y)B(z + y)\,dz\,dy < \infty.$$

(25) is satisfied if the spectrum of B is purely discrete, but generally is violated if any continuous spectrum exists. Hence, the integrals in (24)

diverge in most cases. Thus, (24) is the formal solution of (19) and its derivatives yield the exact solution of (5), but (24) itself ordinarily cannot be interpreted in a pointwise sense.

On the other hand, the exact solution of (5) on the finite interval is still unknown. In this case, there is no known integral equation, but $\tilde{F}(x, t)$ defined by an expansion similar to (24) will still satisfy (19) and converge. Hence, it appears that this procedure might yield the exact solution of (5) on the finite interval.

Further, the method suggests that N-"soliton" solutions of (1) exist for any $n \geqq 2$, and that equations comparable to (19) are the easiest places to find them. The author is not aware of any physical significance of these equations and has not worked out all the details.

In conclusion, it is remarkable that the Korteweg–de Vries equation has an infinite set of conservation laws, that it can be solved exactly, and that two facts are interrelated. However, as shown here, it can almost be solved by a method that works for Burgers' equation, which has one known conservation law. Hence, one is led to speculate that either: (1) the conservation laws are less important than previously suggested, or (2) Burgers' equation also has conservation laws, which have not been discovered.

REFERENCES

1. J. D. Cole, Quart. Appl. Math. **9** (1951), 225–236. MR**13**, 178.
2. C. S. Gardner, J. M. Greene, M. D. Kruskal and R. M. Miura, Phys. Rev. Lett. **19** (1967), 1095.
3. R. Hirota, Phys. Rev. Lett. **27** (1971), 1192.
4. E. Hopf, Comm. Pure Appl. Math. **3** (1950), 201–230. MR**13**, 846.
5. H. Segur, J. Fluid Mech. **59** (1973), 721.
6. G. B. Whitham, Unpublished lecture notes, 1971.

CLARKSON COLLEGE OF TECHNOLOGY

Abstracts

Lectures in Applied Mathematics
Volume 15, 1974

Fixed Point Theorems for Fréchet Spaces and the Existence of Solitary Waves

Jerry Bona and Deb Bose

We examine the question of the existence of solitary wave solutions to simple one-dimensional models for long waves in nonlinear dispersive systems. The models in question take one of the forms:

(1a) $$u_t + u_x + uu_x + Lu_t = 0,$$

or

(1b) $$u_t + u_x + uu_x - Lu_x = 0,$$

where $u = u(x, t)$ is defined on $(-\infty, \infty) \times [0, \infty)$ and L is a linear operator determined by the approximation made to the dispersion relation for the system at hand. (See [**4**, Appendix A] and [**5**].) In particular, if $L = -\partial_x^2$, (1b) is just the well-known Korteweg–de Vries (KdV) equation [**8**], first written down in 1895 as an approximate model for long water waves in a channel, and (1a) is the alternative model whose mathematical properties have been investigated recently ([**4**], [**6**]). Generally, if we use circumflexes to denote Fourier transforms, L is defined by

(2) $$\widehat{(Lu)}(k) = \alpha(k)\hat{u}(k).$$

Thus L is simply a multiplication operator in the transformed plane, or

---

AMS (MOS) subject classifications (1970). Primary 76B25; Secondary 35R20, 35S99, 45G05, 46A40, 47H10, 55C25, 76C10, 86A05.

Key words and phrases. Nonlinear waves, dispersive waves, solitary waves, cones in function space, Fréchet spaces, nonlinear operator equations, compact operators, fixed point theorems, infinite-dimensional degree theory.

in the terminology of modern analysis, a (constant-coefficient) pseudo-differential operator [**13**]. There are physical systems where the appropriate *symbol*, as α is sometimes called, is not a polynomial (in particular, not k^2) and in this case the operator $\boldsymbol{L}$ is not a differential operator. For example, Pritchard [**14**] has investigated experimentally a rotating fluid system where the approximate long-wave model in the forms (1) has the symbol

$$\alpha(k) = k^2\{A + K_0(B|k|)\},$$

where $A, B > 0$ and K_0 is the modified Bessel function of order zero; and Benjamin has worked out a theory of internal waves in the deep ocean, where $\alpha(k) = |k|$.

A question which naturally occurs is whether or not the model systems (1) generally have solitary wave solutions. That is, we ask for a solution u of (1) in the special form

$$u(x, t) = \phi(x - Ct), \tag{3}$$

where C is a constant wave-speed and ϕ is even, nonnegative, and tends monotonically to 0 at $\pm\infty$. Such waves have been a subject of study since Scott Russell's report on waves in 1844. (See [**1**], [**3**], [**8**], [**11**], [**14**] and [**15**].) The KdV equation itself certainly possesses such a solution, which models real solitary waves in a channel of water quite well. Other nonlinear dispersive systems can have solitary waves, as the experimental work of Pritchard has shown. We show that these model equations do have solitary wave solutions for a large variety of symbols.

The method we employ is first to convert the equation for ϕ, obtained by substituting the desired form (3) into (1), into an integral equation. After inverting the operator $\boldsymbol{I} + \boldsymbol{L}$, and scaling out inessential constants, we have the nonlinear convolution equation

$$\phi(x) = \int_{-\infty}^{\infty} K(x - y)\phi^2(y)\, dy = \boldsymbol{A}\phi. \tag{4}$$

In this form, the problem seems naturally amenable to the positive operator methods of Krasnosel'skiĭ ([**9**], [**10**]). The troublesome fact is that we must consider functions defined on the entire real line, in which case it appears that the operator $\boldsymbol{A}$ is not a compact operator in any of the usual Banach function spaces.

This problem is circumvented by considering the operator to act in a (partially ordered) Fréchet space, where it is a compact operator in a suitable sense. Extensions of Krasnosel'skiĭ's positive operator results are proved for a Fréchet space setting using Nagumo's version of Leray–Schauder degree theory [**12**], Dugundji's extension theorem [7], and recent methods due to Benjamin [**2**, Appendix A]. These results are then applied to the solitary wave problem in the form (3), together with a further degree-theory argument, to show existence of the desired waveform.

References

1. T. B. Benjamin, *Internal waves of permanent form in fluids of great depth*, J. Fluid Mech. **29** (1967), 559.

2. ———, *A unified theory of conjugate flows*, Philos. Trans. Roy. Soc. London A **269** (1971), 587.

3. ———, *The stability of solitary waves*, Proc. Roy. Soc. London Ser. A **328** (1972), 153.

4. T. B. Benjamin, J. Bona and J. J. Mahony, *Model equations for long waves in nonlinear dispersive systems*, Philos. Trans. Roy. Soc. London A **272** (1972), 47.

5. T. B. Benjamin and J. Bona, *Model equations for long waves in nonlinear dispersive systems*. II (in preparation).

6. J. Bona and P. J. Bryant, *A mathematical model for long waves generated by wavemakers in nonlinear dispersive systems*, Proc. Cambridge Philos. Soc. **73** (1973), 391–405.

7. J. Dugundji, *An extension of Tietze's theorem*, Pacific J. Math. **1** (1951), 353–367. MR**13**, 373.

8. D. J. Korteweg and G. de Vries, *On the change in form of long waves advancing in a rectangular canal and on a new type of long stationary waves*, Philos. Mag. **39** (1895), 422.

9. M. A. Krasnosel'skiĭ, *Topological methods in the theory of nonlinear integral equations*, GITTL, Moscow, 1956; English transl., Macmillan, New York, 1964. MR**20** #3464; MR**28** #2414.

10. ———, *Positive solutions of operator equations*, Fizmatgiz, Moscow, 1962; English transl., Noordhoff, Groningen, 1964. MR**26** #2862; MR**31** #6107.

11. P. D. Lax, *Integrals of nonlinear equations of evolution and solitary waves*, Comm. Pure Appl. Math. **21** (1968), 467–490. MR**38** #3620.

12. M. Nagumo, *Degree of mapping in locally convex linear topological spaces*, Amer. J. Math. **73** (1951), 485–496. MR**13**, 150.

13. L. Nirenberg, *Pseudo-differential operators*, Proc. Sympos. Pure Math., vol. 16, Amer. Math. Soc., Providence, R.I., 1970, pp. 149–167. MR**42** #5108.

14. W. G. Pritchard, *Solitary waves in rotating fluid*, J. Fluid Mech. **42** (1970), 61.

15. J. Scott Russell, *Report on waves*, Rep. Fourteenth Meeting of the British Assoc., John Murray, London, 1844, p. 311.

Fluid Mechanics Research Institute, University of Essex

Lectures in Applied Mathematics
Volume 15, 1974

Existence of Solutions to the Korteweg–de Vries Initial Value Problem

Jerry Bona and Ronald Smith

For the initial value problem

$$(1) \qquad u_t + uu_x + u_{xxx} = 0, \qquad u(x, 0) = g(x) \qquad \text{for } x \in \boldsymbol{R}, 0 \leqq t \leqq T,$$

we prove the following two theorems:

THEOREM 1. *For $g(x)$ in the space H^s with $s \geqq 3$ there exists a function $u(x, t)$ in $\mathscr{H}^s_T$ which satisfies* (1) (*i.e. the equation is satisfied pointwise almost everywhere*).

THEOREM 2. *The solution of* (1) *depends continuously, in $\mathscr{H}^s_T$, upon perturbations of the initial data in H^s* (*again $s \geqq 3$*).

Here H^s denotes the space of functions with s square integrable derivatives, and $\mathscr{H}^s_T$ denotes the space of functions which evolve continuously in H_s for $0 \leqq t \leqq T$.

The method of proof is to regularize the KdV equation by the addition of a term $-\varepsilon u_{xxt}$, establish results for the regularized problem and then pass to the limit. We remark that the transformation

$$v(X, \tau) = \varepsilon u(\varepsilon^{1/2}(X - \tau), \varepsilon^{3/2}\tau)$$

sends the regularized equation into the BBM initial value problem

$$(2) \qquad v_\tau + v_X + vv_X - v_{XX\tau} = 0, \qquad v(X, 0) = \varepsilon g(\varepsilon^{1/2}X).$$

AMS (MOS) subject classifications (1970). Primary 35K99, 76B15; Secondary 35Q99, 78A40, 86A05.

Key words and phrases. Korteweg–de Vries equation, nonlinear evolution equation, initial value problem, regularization.

The BBM equation (2) was originally proposed as being a rational alternative to the KdV equation (1) for long waves of small but finite amplitude. The results of this paper show that the solutions of these two equations can be made arbitrarily close together if the initial data is in H^3 and corresponds to a sufficiently long and flat wave. Thus in getting a priori estimates or computing solutions it becomes merely a matter of expediency as to which equation is used.

FLUID MECHANICS RESEARCH INSTITUTE, UNIVERSITY OF ESSEX

Lectures in Applied Mathematics
Volume 15, 1974

Entrainment of Frequency in Evolution Equations

James P. Fink and William S. Hall

The existence of periodic solutions near resonance is discussed using elementary methods for the evolution equation

$$\dot{u} = Au + \varepsilon f(t, u)$$

when the linear problem is totally degenerate ($e^{2\pi A} = I$) and the period of f is entrained with ε ($T = 2\pi(1 + \varepsilon\mu)$).

The approach is to solve the periodicity equation $u(T, p, \varepsilon) = p$ for an element $p(\varepsilon)$ in D, the domain of A, as a perturbation from an approximate solution p_0. p_0 is a solution of the nonlinear boundary value problem

$$2\pi\mu Ap + \int_0^{2\pi} e^{-As} f(s, e^{As}p)\, ds = 0$$

obtained from the periodicity equation by dividing by ε, applying the entrainment assumption, and letting $\varepsilon \to 0$. Once p_0 is known, the conventional inverse function theorem is applied in a slightly unconventional form.

Two particular cases where results are obtained are

$$u_t = u_x + \varepsilon\{g(u) - h(t, x)\}$$

with g strongly monotone and

$$\frac{d}{dt}\begin{bmatrix} v \\ w \end{bmatrix} = \begin{bmatrix} 0 & d/dx \\ d/dx & 0 \end{bmatrix}\begin{bmatrix} v \\ w \end{bmatrix} + \varepsilon\begin{bmatrix} v^3 \\ h(t, x) \end{bmatrix},$$

where in both cases D is a certain class of 2π-periodic functions of x.

AMS (MOS) subject classifications (1970). Primary 35B10.

REFERENCES

F. E. Browder, *Existence of periodic solutions for nonlinear equations of evolution*, Proc. Nat. Acad. Sci., U.S.A. **53** (1965), 1100–1103. MR**31** #1558.

L. Cesari, *Existence in the large or periodic solutions of hyperbolic partial differential equations*, Arch. Rational Mech. Anal. **20** (1965), 170–190. MR**32** #1452.

J. Cronin, *Fixed points and topological degree in nonlinear analysis*, Math. Surveys, no. 11, Amer. Math. Soc., Providence, R.I., 1964. MR**29** #1400.

L. di Simone and G. Torelli, *Soluzioni periodiche di equazioni a derivate parziali di tipo iperbolico non lineari*, Rend. Sem. Mat. Univ. Padova **40** (1968), 380–401. MR**37** #4415.

K. O. Friedrichs, *Advanced ordinary differential equations*, Gordon and Breach, New York, 1965. MR**37** #483.

J. K. Hale, *Ordinary differential equations*, Wiley, New York, 1969.

———, *Oscillations in nonlinear systems*, McGraw-Hill, New York, 1963. MR**27** #401.

———, *Periodic solutions of a class of hyperbolic equations containing a small parameter*, Arch. Rational Mech. Anal. **23** (1967), 380–398. MR**34** #6321.

W. S. Hall, *Periodic solutions of a class of weakly nonlinear evolution equations*, Arch. Rational Mech. Anal. **39** (1970), 294–322. MR**43** #672.

P. H. Rabinowitz, *Time periodic solutions of nonlinear wave equations*, Manuscripta Math. **5** (1971), 165–194.

———, *Periodic solutions of nonlinear hyperbolic partial differential equations*, Comm. Pure Appl. Math. **20** (1967), 145–205. MR**34** #6325.

J. J. Stoker, *Nonlinear vibrations in mechanical and electrical systems*, Interscience, New York, 1950. MR**11**, 666.

G. Torelli, *Soluzioni periodiche dell'equazione non lineare* $u_{tt} - u_{xx} + \varepsilon F(x, t, u) = 0$, Rend. Ist. Mat. Univ. Trieste **1** (1969), 123–137. MR**42** #6403.

O. Vejvoda, *The mixed problem and periodic solutions for a linear and weakly nonlinear wave equation in one dimension*, Rozpravy Ceskoslovenske Academic Ved, Rocnik **80** (1970), 1–78.

———, *Periodic solutions of a linear and weakly nonlinear wave equation in one dimension*. I, Czechoslovak Math. J. **14 (89)** (1964), 341–382. MR**30** #5063.

UNIVERSITY OF PITTSBURGH

Lectures in Applied Mathematics
Volume 15, 1974

Perturbation Expansions for Some Nonlinear Wave Equations

James P. Fink, William S. Hall and Siamak Khalili

Perturbation expansions for solutions and period are obtained for the autonomous wave equation

$$Lu = u_{tt} - u_{xx} = f(u),$$

satisfying the periodicity condition

$$u(t + 2\pi/\omega, x) = u(t, x)$$

and the Dirichlet boundary values

$$u(t, 0) = u(t, \pi) = 0.$$

The time scaling $t \to \omega t$, where $\omega = 1 + \varepsilon\mu$, reduces the problem to the form

$$Lu = \varepsilon(-\mu u_{tt} + f(u))$$

where solutions 2π-periodic in t are sought. This can now be treated, at least formally (but not rigorously—yet), as a bifurcation phenomenon with the splitting taking place from the manifold of traveling wave solutions to $Lu = 0$.

Using an orthogonality condition given by Jack Hale, the bifurcation equations reduce to second order, nonlinear, functional differential equations in an unknown initial condition and the unknown parameter μ.

AMS (*MOS*) *subject classifications* (1970). Primary 35C10.

In the special cases where $f(u)$ is given by

$$f(u) = u^3,$$

$$f(u) = -\varepsilon(u + \beta u^3),$$

$$f(u) = -M^2 \sin u \quad \text{(sine–Gordon equation)},$$

the bifurcation equations can be solved explicitly in terms of elliptic functions, and the period determined in terms of elliptic integrals of the first and second kind.

References

1. M. Abramowitz and I. A. Stegun (Editors), *Handbook of mathematical functions, with formulas, graphs and mathematical tables*, Dover, New York, 1966. MR**34** #8606.

2. H. T. Davis, *Introduction to nonlinear differential and integral equations*, Dover, New York, 1962.

3. K. O. Friedrichs, *Lectures on advanced ordinary differential equations*, Gordon and Breach, New York, 1965. MR**37** #483.

4. J. K. Hale, *Periodic solutions of a class of hyperbolic equations containing a small parameter*, Arch. Rational Mech. Anal. **23** (1966), 380–398. MR**34** #6321.

5. ———, *Oscillations in nonlinear systems*, McGraw-Hill, New York, 1963. MR**27** #401.

6. W. S. Hall, *Periodic solutions of a class of weakly nonlinear evolution equations*, Arch. Rational Mech. Anal. **39** (1970), 294–322. MR**43** #672.

7. J. B. Keller and Lu Ting, *Periodic vibrations of systems governed by nonlinear partial differential equations*, Comm. Pure Appl. Math. **19** (1966), 371–420. MR**34** #5347.

8. M. H. Millman and J. B. Keller, *Perturbation theory of nonlinear boundary-value problems*, J. Mathematical Phys. **10** (1969), 342–361. MR**38** #6146.

9. P. H. Rabinowitz, *Periodic solutions of nonlinear hyperbolic partial differential equations*, Comm. Pure Appl. Math. **20** (1967), 145–205. MR**34** #6325.

10. J. Rubinstein, *Sine–Gordon equation*, J. Mathematical Phys. **11** (1970), 258–266. MR**41** #4958.

11. J. J. Stoker, *Oscillations périodiques des systèmes non linéaires ayant une infinité de degrés de liberté, Periodic oscillations of nonlinear systems with infinitely many degrees of freedom*, Actes Colloq. Internat. des Vibrations non Linéaires (Ile de Porquerolles, 1951), Publ. Sci. Tech. Ministère de l'Air, Paris, no. 281, 1953, pp. 61–74. MR**15**, 313.

12. ———, *Nonlinear vibrations in mechanical and electrical systems*, Interscience, New York, 1950. MR**11**, 666.

University of Pittsburgh

Lectures in Applied Mathematics
Volume 15, 1974

An Alternative Derivation of Whitham's Exact Averaged Variational Principle

A. D. Gilbert

1. Background. The material sketched below follows the work of Whitham [**1965a**] concerning the calculation of approximate equations governing the variation of wave parameters (amplitude, wave number and frequency), where it is supposed that these parameters only change on a scale much larger than the scale defined by the wave itself.

Much of Whitham's work ([**1965b**] and subsequently) is concerned with Euler equations from a suitable variational principle. This unifies the theory but is not essential. It leads to the concept of the averaged Lagrangian.

In order to explain the origin of Whitham's approximate equations, Luke [**1966**] studied a particular approximation scheme, suitable for the description of these waves. This used the transformation

$$\psi(x) \to \psi(\varepsilon x, \Theta(\varepsilon x)/\varepsilon), \qquad \frac{d\psi}{dx} \to \varepsilon\frac{\partial\psi}{\partial\varepsilon x} + \varepsilon\frac{d\Theta}{d\varepsilon x}\frac{\partial\psi}{\partial\Theta}, \tag{1}$$

where $\partial\psi/\partial\theta$ denotes the derivative of ψ with respect to its second argument. Here ε is a parameter measuring wave scale to long scale and is $\ll 1$. Θ/ε is a phase function for the wave and ψ has a periodic dependence on phase, while the εx dependence in ψ and Θ allows for the slow variations. It was further found convenient to think of Θ/ε as an independent variable termed θ. Then the transformation above becomes

$$\psi(x) \to \psi(\varepsilon x, \theta), \qquad \frac{d\psi}{dx} \to \varepsilon\frac{\partial\psi}{\partial\varepsilon x} + \frac{d\Theta}{d\varepsilon x}\frac{\partial\psi}{\partial\theta}. \tag{2}$$

AMS (MOS) subject classifications (1970). Primary 41A60, 35G20, 34E20.

In the approximation scheme that followed from this transformation, Whitham's equations were found to be the zero order equations, and all higher order equations could in principle be calculated. The origin of the averaged Lagrangian was not explained.

Whitham [**1970**], using the transformation (2), examined a system governed by a variational principle and obtained, without approximation, a new variational principle, as follows. Starting with

$$\delta \int_D L\left(\psi, \frac{d\psi}{dx}, \varepsilon x\right) dx = 0 \tag{3}$$

there follows, for variations vanishing on the boundary,

$$\frac{d}{dx}\left(\frac{\partial L}{\partial \psi/x}\right) - \frac{\partial L}{\partial \psi} = 0. \tag{4}$$

Under (2), the Lagrangian transforms to

$$L\left(\psi, \varepsilon\frac{\partial \psi}{\partial \varepsilon x} + \frac{d\theta}{d\varepsilon x}\frac{\partial \psi}{\partial \theta}, \varepsilon x\right) \tag{5}$$

and the Euler equation (4) becomes

$$\left\{\varepsilon\frac{\partial}{\partial \varepsilon x} + \frac{d\Theta}{d\varepsilon x}\frac{\partial}{\partial \theta}\right\}\left(\frac{\partial L}{\partial\{\varepsilon\partial\psi/\partial\varepsilon x + (d\Theta/d\varepsilon x)(\partial\psi/\partial\theta)\}}\right) - \frac{\partial L}{\partial \psi} = 0 \tag{6}$$

which may be recognized as being derivable from the variational principle

$$\delta \int_{D_\varepsilon} d(\varepsilon x) \int_a^b L\left(\psi, \varepsilon\frac{\partial \psi}{\partial \varepsilon x} + \frac{d\Theta}{d\varepsilon x}\frac{\partial \psi}{\partial \theta}, \varepsilon x\right) d\theta = 0, \tag{7}$$

for variations which vanish on $B_\varepsilon \times [a, b]$, where B_ε is the boundary of D_ε which is just D scaled by ε, while ψ is a periodic function of θ, with period $b - a$. Equation (7) is the exact averaged variational principle (EAVP), and all results may be explained in terms of it.

2. Sketches of alternative derivations of the EAVP. The derivation of (7) via the Euler equations is rather indirect. We seek a direct way.

(i) Starting with (3), we may replace it exactly by

$$\delta \int_{D_\varepsilon} d\varepsilon x \left\{L\left(\psi, \varepsilon\frac{\partial \psi}{\partial \varepsilon x} + \frac{d\Theta}{d\varepsilon x}\frac{\partial \psi}{\partial \theta}, \varepsilon x\right)\right\}_{\theta = \Theta(\varepsilon x)/\varepsilon} = 0 \tag{8}$$

which makes no further assumptions on whether θ is an independent variable or not. Now in order to obtain (7) from (8), the variational principle (8) must be extended off the line $\theta = \Theta/\varepsilon$ to the whole of $\bar{D} =$

$D_\varepsilon \times [a, b]$. The natural way to do this is to *embed* the variational principle into $\bar{D}$ by supposing that it holds along all lines parallel to $\theta = \Theta/\varepsilon$ in $\bar{D}$. Equation (7) then follows on extending the boundary variations suitably.

(ii) We give a second method for comparison with the Fourier series method employed by Whitham [**1970**]. Let $\{U_n(\theta)\}$ be a complete orthonormal set of functions, periodic in θ, with period $b - a$, and with weight function ρ. As before, (8) follows from (3), and in terms of $\{U_n\}$, equation (8) may be written

$$\sum_n \delta \int_{\bar{D}} d\varepsilon x \int L\left(\psi, \varepsilon\frac{\partial \psi}{\partial \varepsilon x} + \frac{\partial \Theta}{d\varepsilon x}\frac{\partial \psi}{\partial \theta}, \varepsilon x\right) \cdot U_n\left(\theta - \frac{\Theta}{\varepsilon}\right) U_n(0)\rho\left(\theta - \frac{\Theta}{\varepsilon}\right) d\theta = 0. \tag{9}$$

If we choose to set one term of the series (9) separately to zero, it may be shown that all the other terms separately vanish, and then (7) is easily constructed using the separate vanishing of each term in (9).

3. Further points. The equivalence of the embedding in method (i) and the selection of the separate vanishing of a single term in method (ii) is easily demonstrated.

We note that this simple embedding can take place, because the transformed equation (6), which was initially only considered on the line $\theta = \Theta/\varepsilon$, is invariant under translation of θ.

REFERENCES

J. C. Luke, Proc. Roy. Soc. Ser. A **292** (1966), 403–412. MR**33** #3491.
G. B. Whitham, Proc. Roy. Soc. Ser. A **283** (1965a), 238–261. MR**31** # 996.
———, J. Fluid Mech. **22** (1965b), 273–283. MR**31** #6459.
———, J. Fluid Mech. **44** (1970), 373–395. MR**42** #4858.

DEPARTMENT OF APPLIED MATHEMATICS AND THEORETICAL PHYSICS, UNIVERSITY OF CAMBRIDGE

Lectures in Applied Mathematics
Volume 15, 1974

WKB–KdV

John M. Greene

There is some pedagogical interest in applying the various versions of WKB method to the Korteweg–de Vries equation. The fast, oscillatory behavior here consists of cnoidal waves depending on three adjustable parameters. Thus this equation provides a rich, yet analytically tractable, testing ground. Calculations of this type have been given by Whitham [**1**] and by Miura and Kruskal [**2**].

Following Miura and Kruskal, we assume a solution of the form

$$U = U(\theta, x, t, \delta), \qquad \theta \equiv \frac{B(x, t)}{\delta},$$

where δ is an expansion parameter. In lowest order we obtain

$$(B_t/B_x + U)U_\theta + B_x^2 U_{\theta\theta\theta} = 0,$$

with solution

$$U = (\alpha - \gamma)\left\{\frac{1 + k^2}{3} - k^2 \mathrm{Sn}^2\left[\left(\frac{\alpha - \gamma}{12B_x^2}\right)^{1/2}(\theta - \theta_0), k\right] - B_t/B_x\right\}.$$

The periodicity condition, $U(\theta + 1) = U(\theta)$, yields

$$(\alpha - \gamma)/12B_x^2 = 4K^2(k).$$

On the long time scale, the slowly varying quantities can be taken to be $\alpha - \gamma$, k^2 and B_t/B_x. The latter quantity, together with the value for B_x from above, determines the slow variation of θ. Thus the third slowly

---

AMS (MOS) subject classifications (1970). Primary 76B25, 35B25.

varying quantity has a slightly different character from the other two, being related indirectly to an integration constant through the periodicity condition.

Higher order equations yield the homogeneous operator

$$\left[\frac{B_t}{B_x}\frac{\partial}{\partial\theta} + \frac{\partial}{\partial\theta}U^{(0)} + B_x^2\frac{\partial^3}{\partial\theta^3}\right].$$

Equations for the slowly varying quantities can be found by finding annihilators of this operator and applying them to the original Korteweg–de Vries equation. Two such annihilators are

$$\oint d\theta\Big\{, \qquad \oint d\theta U\Big\{.$$

A third involves nonperiodic functions and is not useful. Instead, we consider

$$\oint d\theta(U^2 + 2B_x^2 U_{\theta\theta})\Big\{,$$

which is not independent of the previous two in lowest order, but can be given content by combining it with the consistency relation

$$(B_x)_t = [B_x(B_t/B_x)]_x.$$

Essentially this procedure was followed in both the above references. Here again, the third slowly varying quantity has a different character from the other two.

It would be more satisfying if, in some deeper sense, the three variables of the theory had a uniform foundation.

References

1. G. B. Whitham, *Non-linear dispersive waves*, Proc. Roy. Soc. Ser. A **283** (1965), 238–261. MR**31** #996.

2. R. M. Miura and M. D. Kruskal, *Application of a nonlinear WKB method to the Korteweg–de Vries equation*, SIAM. J. Appl. Math. **26** (1974) (to appear).

Princeton University

Lectures in Applied Mathematics
Volume 15, 1974

Model Equations for Wavy Viscous Film Flow

G. M. Homsy

For some time considerable theoretical interest has accrued in the description of wave formation in thin films of viscous fluids. The simplest example of flows of this class is that of the flow of a single fluid down inclined or vertical planes. The decisive treatment of the linear stability of the simple rectilinear Nusselt solution is due to Benjamin [**3**] and was further clarified by Yih [**14**]. The most unstable mode in this case is a long interfacial wave which is unconditionally unstable if the plane is vertical. Since this work, many other flows have been demonstrated to exhibit this mode of instability, e.g. [**7**], [**9**], [**15**].

With the linear stability firmly understood, attention is properly turned to an understanding of the process of wave evolution and equilibration under linearly unstable conditions. Much work has been done along these lines by Lin ([**10**], [**11**], [**12**]), who applied the Stuart–Watson method to "free surface" flows. However, this method finds its greatest utility only near the neutral curve, thus excluding the consideration of the wave number of maximum growth rate. We have therefore chosen to treat the problem using Benney's [**4**] long wave expansions. This approach has the advantage of solving the dynamical problem *exactly* in terms of the unknown surface elevation $h(x, t)$. We take the motion to be two dimensional (but see [**13**]). Thus in terms of the stream function, defined such that $(u, v) = (\psi_y, -\psi_x)$ and stretching (x, t) by a long wave parameter α, we have dimensionlessly

$$\nabla_\alpha^4 \psi = \alpha \operatorname{Re} (\partial \nabla_\alpha^2 \psi / \partial t + \partial(\psi, \nabla_\alpha^2 \psi)/\partial(y, x)), \tag{1a}$$

AMS (MOS) subject classifications (1970). Primary 76D05, 76E05, 76E30.

(1b) $$\nabla_\alpha^2 = \partial^2/\partial y^2 + \alpha^2\partial^2/\partial x^2.$$

We seek a solution as an expansion in α, viz.

(2) $$\psi = \sum_n \psi_n(y; h, h_x, \dots)\alpha^n$$

subject to the dynamical conditions of vanishing tangential stress and balancing of normal stress at the unknown interface and the no-slip condition at the solid boundary. These conditions have been given conveniently in terms of ψ by Krantz and Goren [8]. It is clear from equation (1) that the expansion will result in each ψ_n being the solution to

$$\partial^4\psi_n/\partial y^4 = g_n(y, h, h_x, \dots), \qquad g_0 = 0,$$

and hence each ψ_n will be a polynomial in y with coefficients which depend upon h and its derivatives.

The evolution equation is obtained from the kinematic condition

(3a) $$Dh/Dt = v, \qquad y = h(x, t),$$

or

(3b) $$\frac{\partial h}{\partial t} + \frac{\partial\psi}{\partial y}\frac{\partial h}{\partial x} + \frac{\partial\psi}{\partial x} = 0, \qquad y = h(x, t),$$

which has the conservation form

(3c) $$\partial h/\partial t + \{\psi(y; h, h_x, \dots)|_{y=h}\}_x = 0$$

expressing the fact that the interface is a material surface.

As noted by Gjevik [5] and Krantz and Goren [8] the lowest order evolution equation which can equilibrate is

(4) $$h_t + (h^3)_x + \alpha \operatorname{Re}\left\{\frac{6}{5}\left(\frac{\operatorname{Re} h^3 - \operatorname{Re}_c}{\operatorname{Re}}\right)h^3 h_x + \frac{P}{3}h^3 h_{xxx}\right\}_x = 0.$$

Here $\operatorname{Re}_c = (5/6)\cot\beta$, $P = \alpha^2 We$ is a surface tension parameter which equals 9/5 if α is chosen as the wave number of maximum growth rate according to linear theory. Neglect of this term leads to more rapid growth in the nonlinear range [4] and a forward-breaking wave, since at a crest $h > 1$, and hence $h^3 \operatorname{Re} > \operatorname{Re}_c$. As it stands, equation (4) is too general, since it describes long waves of arbitrary amplitude. It may be shown [2] that the wave that equilibrates has amplitude of $O(\alpha \operatorname{Re})$ and hence if we set $h = 1 + \alpha \operatorname{Re}\eta$ we find

(5) $$\eta_t + 3\eta_x + \alpha \operatorname{Re}\left\{3(\eta^2)_x + \frac{6}{5}\left(\frac{\operatorname{Re} - \operatorname{Re}_c}{\operatorname{Re}}\right)\eta_{xx} + \frac{P}{3}\eta_{xxxx}\right\} = 0.$$

Equation (5) is the quintessential *model* equation for this class of long interfacial waves. Numerical studies of (5) are underway.

Returning to equation (3c), we can seek steady waves of the form $h(x, t) = h(x - ct)$ and we have

$$-ch' + \psi'(h; h, h_x, \dots) = 0 \tag{6a}$$

which has a trivial first integral

$$-ch + \psi(h; h, h_x, \dots) = A \tag{6b}$$

and the analog of equation (5) becomes

$$-(c + 3)\eta' + \alpha \operatorname{Re} \left\{ 6\eta\eta' + \frac{6}{5}\eta'' + \frac{P}{3}\eta'''' \right\} = 0. \tag{6c}$$

Gjevik [**5**] and Atherton and Homsy [**2**] used Stuart-Watson amplitude expansions to treat equation (6a) and found the results to be poorly convergent. With α fixed, these previous treatments have found approximate solutions to equation (6a) which are forced to be 2π-periodic. It is well known, however, that such nonlinear autonomous systems will in general have a period dependent upon amplitude. Numerical experiments indicate that periodic solutions of equation (6c) subject to random initial data have natural period $\sim 1.4\pi$.

We have extended the expansion in α to higher order using REDUCE, a symbolic manipulation language capable of a large variety of operations, symbolic partial differentiation and multiplication and regrouping of power series being especially useful in the present context ([**1**], [**6**]). Scaling arguments [**1**] leads to the conclusion that a linear dispersion term arising from these higher order terms should be added to equation (5) to yield

$$\begin{aligned} \eta_t + 3\eta_x + \alpha \operatorname{Re} \Bigg\{ & 3(\eta^2)_x + \frac{6}{5}\left(\frac{\operatorname{Re} - \operatorname{Re}_c}{\operatorname{Re}}\right)\eta_{xx} \\ & + \frac{P}{3}\eta_{xxxx} + D \operatorname{Re}^{-1/6} \eta_{xxx} \Bigg\} = 0 \end{aligned} \tag{7}$$

where D is an $O(1)$ constant depending only upon fluid properties. It is of interest to note that equations (4), (5), or (7) are *essentially* nonlinear in the sense that the nonlinearity cannot be removed by WKB-type approximations. It is perhaps not surprising then that Stuart–Watson expansions converge poorly.

REFERENCES

1. R. W. Atherton, Engineer's Thesis, Stanford University, Stanford, Calif., 1972.
2. R. W. Atherton and G. M. Homsy, unpublished calculations.

3. T. B. Benjamin, J. Fluid Mech. **2** (1957), 554. MR**24** #B1050.
4. D. J. Benney, J. Mathematical Phys. **45** (1966), 150. MR**34** #1010.
5. B. Gjevik, Phys. Fluids **13** (1970), 1918.
6. A. C. Hearn, *REDUCE 2 users manual*, Stanford Artificial Intelligence Project Memo AIM-133, 1970.
7. C. E. Hickox, Phys. Fluids **14** (1971), 251.
8. W. B. Krantz and S. L. Goren, I. & E. C. Fund. **9** (1970), 107.
9. S. P. Lin, Phys. Fluids **10** (1967), 64.
10. ———, J. Fluid Mech. **36** (1969), 113.
11. ———, J. Fluid Mech. **40** (1970), 307.
12. ———, Phys. Fluids **14** (1971), 263.
13. G. Roskes, Phys. Fluids **13** (1970), 1440.
14. C.-S. Yih, Phys. Fluids **6** (1963), 321.
15. ———, J. Fluid Mech. **27** (1967), 337.

STANFORD UNIVERSITY

Lectures in Applied Mathematics
Volume 15, 1974

Approximating Nonlinear Waves

Gray Jennings

The substance of this note is the announcement of the existence of traveling waves for a class of difference schemes of the form

$$
\begin{aligned}
u_j^{n+1} = u_j^n - (\Delta t/\Delta x)\{&g(u_{j+k}^n, u_{j+k-1}^n, \dots, u_{j-k+1}^n) \\
&- g(u_{j+k-1}^n, u_{j+k-2}^n, \dots, u_{j-k}^n)\}
\end{aligned}
\tag{1}
$$

where $u(j\Delta x, n\Delta t) = u_j^n$, u_j and $g(u_1, u_2, \dots, u_{2k})$ are scalars. The right-hand side of (1) is required to be strictly monotone increasing in each of the variables on which it depends. This condition limits the difference scheme to being no more than first order accurate as an approximation to

$$
\frac{\partial u}{\partial t} + \frac{\partial}{\partial x} f(u) = 0. \tag{2}
$$

The condition $g(u, u, \dots, u) = f(u)$ couples the two relations. Monotone traveling waves exist with limits u_r and u_l at plus and minus infinity if u_r and u_l determine a shock which satisfies the jump condition and Condition E which are respectively

$$
\begin{aligned}
s(u_r - u_l) &= f(u_r) - f(u_l), \\
s(u - u_r) &> f(u) - f(u_r), \quad \text{for } u \in (u_r, u_l).
\end{aligned}
\tag{3}
$$

If $u_l < u_r$, the inequality in (3) is reversed.

AMS (MOS) subject classifications (1970). Primary 39A10, 39L60, 39L65, 39R05; Secondary 35L05, 35D99, 39L40, 76L05.

The sharp transition in the shock has a parallel in the traveling wave of the difference scheme—most of the change from u_r to u_l occurs in 3 to 5 mesh widths. If $u_0(x)$ is an initial condition which is resolved under (2) into a single shock, then an approximation generated by (1) will for sufficiently large time converge to a traveling wave of (1) and the distance between the jump in the continuous problem and the region of rapid transition for the discrete wave will be $K(\Delta x)$ where K is independent of time. Schemes which approximate the propagation of discontinuities by linear equations typically spread the discontinuity.

For a discussion of the mathematics of weak solutions of hyperbolic partial differential equations see [**3**], [**4**], and [**5**]. The essential feature of (1) is that it conserves precisely $\sum_j u_j^n$, the discrete analogue of $\int u(x, t)\,dx$. That the integral of u is conserved under the partial differential equation when discontinuities are allowed follows from the Rankine–Hugoniot relations.

The existence of traveling waves in this case is closely analogous to the work of Foy [**2**] and more recently of Conley and Smoller [**1**] on approximating shocks by solutions of

$$\frac{\partial u}{\partial t} + \frac{\partial}{\partial x} f(u) = \varepsilon \frac{\partial^2}{\partial x^2} u \quad \text{when } \varepsilon > 0.$$

The proof of these results will appear elsewhere. The bibliography contains a sampling of the literature and some of the more recent papers. The reader who is interested in the state of the art should start with [**4**] and [**5**] which contain excellent bibliographies.

References

1. Charles C. Conley and Joel A. Smoller, *Viscosity matrices for two dimensional nonlinear hyperbolic systems*, Comm. Pure Appl. Math. **23** (1970), 867–884. MR**43** #714.

2. Linus R. Foy, *Steady state solutions of hyperbolic systems of conservation laws with viscosity terms*, Comm. Pure Appl. Math. **17** (1964), 177–188. MR**28** #2354.

3. P. D. Lax, *Hyperbolic systems of conservation laws*. II, Comm. Pure Appl. Math. **10** (1957), 537–566. MR**20** #176.

4. T. Nishida, *Global solution of an initial boundary value problem of a quasilinear hyperbolic system*, Proc. Japan Acad. **44** (1968), 642–646. MR**38** #4821.

5. B. Quinn, *Time decreasing functionals of nonlinear conservation laws*, Comm. Pure Appl. Math. **24** (1971), 125–132.

Courant Institute of Mathematical Sciences, New York University

Lectures in Applied Mathematics
Volume 15, 1974

The Korteweg–de Vries Equation with Variable Coefficients

R. S. Johnson

The phenomenon of a single solitary wave moving into a region of decreasing depth and producing solitons is described. In particular the numerical integration of the Euler equations and the experimental data, both given by Madsen and Mei [**1969**], are discussed.

Since the results show many similarities with the standard theory of the KdV equation, it seems reasonable to seek such an equation to govern this behavior. It is shown that the Euler equations may be reduced, by the application of a formal asymptotic procedure, to the following equation:

$$U_X + \tfrac{3}{2} d^{-7/4} U U_\xi + \tfrac{1}{6} K\, d^{1/2} U_{\xi\xi\xi} = 0. \tag{*}$$

This is a variable coefficient KdV equation, where $d(X)$ is the local depth and the surface wave amplitude is proportional to $\varepsilon U(X, \xi)\, d^{-1/4}$. The asymptotic analysis requires $\varepsilon \to 0$, together with

(a) $(\text{depth/wave length})^2 = K\varepsilon$ (long waves),

(b) depth slowly varying: $d = d(\varepsilon x)$, $\varepsilon x = X$,

and (*) is then the equation obtained in the far-field

$$X = \varepsilon x = O(1),$$

$$\xi = \int_0^x d^{-1/2}(\varepsilon x)\, dx - t = O(1).$$

ξ is thus the appropriate linearised characteristic coordinate.

AMS (MOS) subject classifications (1970). Primary 76B25.

Key words and phrases. Korteweg–de Vries equation, water waves, variable depth water, asymptotic solution, numerical solution.

Attention is now turned to various approximate solutions of the equation (*) for a solitary wave initial condition. That is, in the region of constant unit depth, a solitary wave is propagated with unchanging form which then enters a variable depth region. Two particular cases are studied in detail:

(i) Rapid, but smooth, decrease in depth to a new (nonzero) value—a shelf. Although the application of this result to the *physical* problem is somewhat in doubt—we have already assumed a slow change in depth—the final conclusions are convincing. By making use of the infinite sequence of quasi-conservation laws for (*) it is shown that in the limit of a rapid change of depth the initial waveform stays essentially unaltered. Thus on the shelf (of depth $d_0 < 1$) the initial condition is as originally prescribed and consequently the problem is simply the solution of (*) with constant coefficients ($d = d_0$) and solitary wave initial data. Standard KdV theory then predicts that exactly n solitons will eventually appear on the shelf if

$$d_0 = [\tfrac{1}{2}n(n + 1)]^{-4/9}$$

where the soliton amplitudes are $2Am^2/n(n + 1)$, $m = 1, 2, \ldots, n$. [A is the amplitude of the initial solitary wave.] These predictions are confirmed by various integrations of (*), together with the case considered by Madsen and Mei ($d_0 = 0.5$) for the full inviscid problem.

(ii) Very slow decrease in depth to a new (nonzero) value. For this problem, it is now assumed that $d(X)$ is a slowly varying function, i.e. $d = d(\sigma X)$, $\sigma \to 0$. A two-time scale procedure is adopted by seeking a solution as a slowly varying solitary wave. This expansion is shown not to be uniformly valid (cf. Grimshaw [**1970**]) but it may be successfully matched to zero ahead of the leading wave, and to other solitons behind.

In conclusion, it is briefly mentioned that for steadily (slowly) increasing depth a uniformly valid solution can be obtained as a slowly varying cnoidal wave.

References

R. Grimshaw 1970, *The solitary wave in water of variable depth*, J. Fluid Mech. **42**, 639–656.

R. S. Johnson 1973, *On the development of a solitary wave moving over an uneven bottom*, Proc. Cambridge Philos. Soc. **73**, 183–203.

———, 1972, *Some numerical solutions of a variable-coefficient Korteweg–de Vries equation* (*with applications to solitary wave development on a shelf*), J. Fluid Mech. **54**, 81–91.

T. Kakutani 1971, *Effect of an uneven bottom on gravity waves*, J. Phys. Soc. Japan **30**, 272–276.

D. J. Korteweg and G. de Vries 1895, *On the change of form of long waves advancing in a rectangular canal, and on a new type of long stationary waves*, Philos. Mag. **39**, 422–433.

O. S. Madsen and C. C. Mei 1969, *The transformation of a solitary wave over an uneven bottom*, J. Fluid Mech. **39**, 781–791.

H. Ono 1972, *Wave propagation in an inhomogeneous anharmonic lattice*, J. Phys. Soc. Japan **32**, 332–336.

F. D. Tappert and N. J. Zabusky 1971, *Gradient-induced fission of solitons*, Phys. Rev. Lett. **27**, 1774–1776.

UNIVERSITY OF NEWCASTLE UPON TYNE

Lectures in Applied Mathematics
Volume 15, 1974

Landform Evolution Models

Jon C. Luke

Following the early qualitative work by Penck [**1924**] on differential erosion of landforms, several authors have studied quantitative models. For example, Scheidegger [**1961**] and Young [**1963**] give results of numerical work, and Mitin and Trofimov [**1964**] obtain explicit algebraic solutions for a few special cases. Scheidegger [**1970**] gives a survey of the related literature.

If the equation $h_t = -f(h_x, h_y)$ is taken as a simple model for erosion of land surfaces, it is easy to work out examples graphically (Luke [**1972**]) by the method of characteristics. As in many nonlinear problems, shock solutions develop; here they occur as sharp corners, that is, as jumps in h_x and h_y. Stratification and explicit effects of fluid flow can be incorporated in more complicated models. By means of the substitution $u = h_x$, the one-dimensional problem $h_t = -f(h_x)$ may be shown equivalent to the kinematic wave equation $u_t + f'(u)u_x = 0$, which was used by Lighthill and Whitham [**1955**] in the study of traffic flow.

References

M. J. Lighthill and G. B. Whitham, *On kinematic waves*. I. *Flood movement in long rivers*, Proc. Roy. Soc. London. Ser. A **229** (1955), 281–316. MR**17**, 309.

J. C. Luke, *Mathematical models for landform evolution*, J. Geophys. Res. **77** (1972), 2460.

A. V. Mitin and A. M. Trofimov, *On a generalization of the theory of Bakker and Scheidegger*, Uč. Zap. Kazan. Gos. Univ. **124** (1964), 112. (Russian)

AMS (*MOS*) *subject classifications* (1970). Primary 73Q05; Secondary 35A30.

W. Penck, *Die morphologische Analyse*, 1924; English transl., *Morphological analysis of land forms*, Macmillan, London, 1953.

A. Scheidegger, *Mathematical models of slope development*, Bull. Geol. Soc. Amer. **72** (1961), 37.

———, *Theoretical geomorphology*, 2nd ed., Springer, Berlin, 1970.

A. Young, *Deductive models of slope evolution*, Nachr. Akad. Wiss. Goettingen. Math. Phys. Kl. **1963**, no. 5, 45.

UNIVERSITY OF CALIFORNIA, SAN DIEGO

Lectures in Applied Mathematics
Volume 15, 1974

Cross Waves

J. J. Mahony

A theory was described for an instability mechanism which could explain the development of cross waves in an open channel. In this work the theory of resonant interactions was extended to deal with localized fields which are only approximately modal. The mathematics involves rather complicated differential integral equations. The theory gives good agreement with some recently obtained experimental results. The problems of extending the theory to deal with the equilibrium state after instability has set in were discussed.

The work has appeared in the following references.

REFERENCES

J. J. Mahony, *Cross waves*, J. Fluid Mech. **55** (1972), 229–244.

B. J. S. Barnard and W. G. Pritchard, *Cross waves*. II: *Experiment*, J. Fluid Mech. **55** (1972), 245–255.

UNIVERSITY OF ESSEX

AMS (MOS) subject classifications (1970). Primary 76B15.
Key words and phrases. Stability.

Lectures in Applied Mathematics
Volume 15, 1974

Validity of Averaging Methods for Certain Systems with Periodic Solutions

J. J. Mahony

Outlines were given of proofs of theorems concerning the validity of solutions obtained by heuristic asymptotic methods for both ordinary and partial differential equations. The motivation was to obtain guidance about the validity of methods commonly used to calculate the development of slowly varying wave trains. The approach assumed that some trial function had been obtained, by means unspecified, for which asymptotic satisfaction of the differential equation could be demonstrated. For the systems considered energy estimates combined with the contraction mapping theorem enable one to establish that any such trial approximation is an asymptotic representation of the exact solution on a time interval on which significant development of the system occurs. It was also shown, by counterexample, that there can be systems where standard approximation methods fail, and the reason for such failure was demonstrated. The validity of averaging methods appears to depend more strongly on the overall stability of the system that seems to have been recognized widely in the literature.

This work has appeared under the above title in the Proc. Roy. Soc. London Ser. A **330** (1972), 349–371.

UNIVERSITY OF ESSEX

AMS (MOS) subject classifications (1970). Primary 34E15, 35L60, 35B40.

Lectures in Applied Mathematics
Volume 15, 1974

The Existence of Unbounded Solutions of the Korteweg–de Vries Equation

A. Menikoff

We consider solutions of the Korteweg–de Vries equation for which $|u| \to \infty$ as $|x| \to \infty$. For this case we give a new proof of the result of Gardner et al. [**1**].

THEOREM. *If* $u_t + uu_x + u_{xxx} = 0$ *and* $u \to -\infty$ *as* $|x| \to \infty$ *then the eigenvalues of*

$$\frac{d^2\psi}{dx^2} + \frac{1}{6}u\psi = \lambda\psi$$

stay fixed as t varies.

By means of a finite difference scheme similar to Sjöberg [**2**] and using the polynomial integrals of the Korteweg–de Vries equation of Miura et al. [**3**] we show that such solutions exist.

THEOREM. *If* $(\partial^n/\partial x^n)(u_0(x)) = O(x^{1-n})$ *as* $|x| \to \infty$, $0 \leqq n \leqq 7$, *then the equation* $u_t + uu_x + u_{xxx} = 0$, $u(x, t) = u_0(x)$ *at* $t = 0$, *has a unique smooth solution.*

For further details see [**4**].

REFERENCES

1. C. Gardner et al., *Method for solving the Korteweg–de Vries equation*, Phys. Rev. Lett. **19** (1967), 1095–1097.

AMS (MOS) subject classifications (1970). Primary 35G20, 35G25; Secondary 34B25.

2. A. Sjöberg, *On the Korteweg–de Vries equation: existence and uniqueness*, J. Math. Anal. Appl. **29** (1970), 569–579.

3. R. Miura et al., *Korteweg–de Vries equation and generalizations.* II. *Existence of conservation laws and constants of motion*, J. Mathematical Phys. **9** (1968), 1204–1209. MR**40** #6042b.

4. A. Menikoff, *The existence of unbound solution of the Korteweg–de Vries equation*, Comm. Pure Appl. Math. **25** (1972), 407–432.

JOHNS HOPKINS UNIVERSITY

Lectures in Applied Mathematics
Volume 15, 1974

Solutions in the Large for Some Nonlinear Hyperbolic Systems of Conservation Laws

Takaaki Nishida and Joel A. Smoller

We are concerned with an existence theorem of weak solutions in the large (in time) of the initial value problem for some nonlinear hyperbolic systems of conservation laws which describes one-dimensional ideal isentropic gas motion:

$$\text{(1)} \quad \partial v/\partial t - \partial u/\partial x = 0, \qquad \partial u/\partial t + \partial p(v)/\partial x = 0 \qquad \text{in } t \geqq 0, x \in R,$$

$$\text{(2)} \quad v(0, x) = v_0(x), \qquad u(0, x) = u_0(x) \qquad \text{in } x \in R.$$

Here t is the time, x is the Lagrangian coordinate, v is the specific volume, u is the velocity of the gas and p is the pressure. The equation of state is $p = k^2/v^\gamma$ (k = const), γ = the ratio of specific heat $= 1 + 2\varepsilon$, and ε is assumed to be a small positive constant in this work.

Since the system (1) is hyperbolic and genuinely nonlinear in the sense of Lax, it is well known that the initial value problem (1), (2) does not generally have a global smooth solution for even smooth initial values (2). We thus define and seek a weak solution for this problem in the sense of the theory of distributions.

THEOREM. *Let the initial values $v_0(x)$, $u_0(x)$ have bounded total variation and be bounded, i.e., $|u_0(x)| \leqq M$, $0 < \underline{v} \leqq v_0(x) \leqq \bar{v} < +\infty$ (M, $\underline{v}$, $\bar{v}$ are some constants), $x \in R$. Then there exists a γ_0, $1 < \gamma_0 < 2$, such that for any adiabatic constant of the gas $\gamma \in [1, \gamma_0]$, the initial value problem* (1), (2) *has a global weak solution which has bounded total variation on each line t = constant > 0.*

AMS (*MOS*) *subject classifications* (1970). Primary 35L65; Secondary 76A99.

Our method involves a detailed consideration of the global properties of the shock-wave curves which enables us to obtain estimates for the interaction of shock- and rarefaction-waves. Two essential cases are the following: In the interaction of two shock-waves of the opposite kind, the shock strengths of those shock-waves increase only by the order of ε. In the interaction of the shock- and rarefaction-waves of the same kind the shock strength decreases.

These estimates are then applied to the Glimm difference scheme.

The existence of global weak solutions for the nonlinear hyperbolic system of conservation laws is considered in the papers [**1**]–[**10**] and others. These papers do not contain our results.

The piston problem for the system (1) may be treated analogously under the same assumption on the initial values and the piston velocity, i.e.,

$$\varepsilon \cdot \{\text{total variation of the initial value and the piston velocity}\}$$

is sufficiently small.

References

1. B. Riemann, *Über die Fortpflanzung ebener Luftwellen von endlicher Schwingungsweite*, Abh. Koenig Ges. Wiss. Göttingen **8** (1860), 43.

2. P. D. Lax, *Hyperbolic systems of conservation laws*. II, Comm. Pure Appl. Math. **10** (1957), 537–566. MR**20** #176.

3. J. Glimm, *Solutions in the large for nonlinear hyperbolic systems of equations*, Comm. Pure Appl. Math. **18** (1965), 697–715. MR**33** #2976.

4. J. Glimm and P. D. Lax, *Decay of solutions of systems of nonlinear hyperbolic conservation laws*, Mem. Amer. Math. Soc. No. 101 (1970). MR**42** #676.

5. T. Nishida, *Global solutions for an initial boundary value problem of a quasilinear hyperbolic system*, Proc. Japan Acad. **44** (1968), 642–646. MR**38** #4821.

6. J. L. Johnson and J. A. Smoller, *Global solutions for an extended class of hyperbolic systems of conservation laws*, Arch. Rational Mech. Anal. **32** (1969), 169–189. MR**38** #4822.

7. N. S. Bahvalov, *On the existence of regular solutions in the large for the quasilinear hyperbolic systems*, Z. Vyčisl. Mat. i Mat. Fiz. **10** (1970), 969–980. (Russian.) MR**43** #5165.

8. N. Borovikov, *On the existence of global solutions for a class of quasilinear hyperbolic systems*, Moskow, 1971 (preprint).

9. R. DiPerna, *Global solutions to a class of nonlinear hyperbolic systems of equations*, Comm. Pure Appl. Math. **26** (1973), 1–28.

10. J. Greenberg, private communication.

11. P. D. Lax, *A concept of entropy*, Contributions to Nonlinear Functional Analysis (E. H. Zarantonello, editor), Academic Press, New York, 1971, pp. 603–634.

Courant Institute of Mathematical Sciences, New York University

University of Michigan

Lectures in Applied Mathematics
Volume 15, 1974

The Sine–Gordon Equation

Alwyn C. Scott

The nonlinear wave equation

$$\phi_{xx} - \phi_{tt} = \sin \phi \tag{1}$$

was introduced in 1939 by Frenkel and Kontorova in connection with the propagation of a "slip" dislocation in a one-dimensional crystal ([**1**], [**2**]). More recently it has been applied to the motion of a Bloch wall between ferromagnetic domains ([**3**], [**4**], [**5**]), to magnetic flux propagation in a large Josephson type superconducting tunnel junction ([**6**], [**7**], [**8**]), and as a model unitary field theory for elementary particles ([**9**], [**10**]). A closely related equation has been used to describe the propagation of a short optical pulse through a resonant laser medium ([**11**], [**12**]). Several reviews have recently been published ([**11**], [**12**], [**13**], [**14**]).

Since (1) is invariant to a Lorentz transformation of the independent variables, traveling wave solutions (of velocity u) can be written from a generalization of the simple pendulum problem as

$$\int^{\phi_T} \frac{d\phi}{(2(E - \cos \phi))^{1/2}} = \frac{\pm(x - ut)}{(1 - u^2)^{1/2}} \tag{2}$$

where E is a constant of integration. Linear perturbation analysis indicates that stable solutions are only obtained for $E \geqq 1$ and $|u| < 1$. For $E = 1$, (2) indicates a *soliton* of the form

$$\phi_T = 4 \tan^{-1} \left[\exp \left(\frac{\pm(x - ut)}{(1 - u^2)^{1/2}} \right) \right]. \tag{3}$$

AMS (MOS) subject classifications (1970). Primary 35L60.

Since the total rotation must be conserved in any collision, the ($\pm$) signs in (3) can be interpreted as indicating a soliton and an "antisoliton." For $E = -1$ and $|u| > 1$, (2) indicates a soliton which travels faster than the characteristic velocity. Such a "tachyon," however, is unstable.

The nondestructive collision of two solitons was reported by Perring and Skyrme in 1962 [**15**]. Although this result was first obtained through computer calculations, they found an *exact* solution of (1) for a soliton–soliton collision to be

$$\phi = 4 \tan^{-1} \left[\frac{u \sinh (x/(1 - u^2)^{1/2})}{\cosh (ut/(1 - u^2)^{1/2})} \right] \tag{4}$$

with a corresponding expression for a soliton–antisoliton collision. Such events can be conveniently demonstrated on a mechanical analog for (1) [**16**].

Solutions which represent an arbitrary number of interacting solitons can be constructed using the formalism of the *Bäcklund transformation* ([**2**], [**12**], [**17**], [**18**]). Each successive application of such a transformation introduces a new soliton, the velocity of which can be considered a new constant of the motion ([**19**], [**20**]).

References

1. J. Frenkel and T. Kontorova, *On the theory of plastic deformation and twinning*, Acad. Sci. J. Phys. USSR **1** (1939), 137–149. MR**1**, 190.

2. A. Kochendörfer, A. Seeger and H. Donth, *Theorie der Versetzungen in eindimensionalen Atomreihen*, Z. Phys. **127** (1950), 533–550; **130** (1951), 321; **134** (1953), 173. MR**12**, 304.

3. W. Döring, *Über die Trägheit der Wände zwischen weisschen Berzirken*, Z. Naturf. **3a** (1948), 373.

4. R. Becker, *La dynamique de la paroi de Bloch et la perméabilite en haute fréquence*, J. Phys. Radium **12** (1951), 332.

5. U. Enz, *Die Dynamik der blochschen Wand*, Helv. Phys. Acta **37** (1964), 245.

6. I. O. Kulik, *Propagation of waves in a Josephson tunnel junction in the presence of vortices and the electrodynamics of weak superconductivity*, Soviet Phys. JETP **24** (1967), 1307.

7. P. Lebwohl and M. J. Stephen, *Properties of vortex lines in superconducting barriers*, Phys. Rev. **163** (1967), 376.

8. A. C. Scott, *Steady propagation on long Josephson junctions*, Bull. Amer. Phys. Soc. **12** (1967), 308; *A nonlinear Klein–Gordon equation*, Amer. J. Phys. **37** (1969), 52.

9. T. H. R. Skyrme, *A nonlinear theory of strong interactions*, Proc. Roy. Soc. A **247** (1958), 260.

10. U. Enz, *Discrete mass, elementary length, and a topological invariant as a consequence of a relativistic invariant variational principle*, Phys. Rev. (2) **131** (1963), 1392–1394. MR**30** #5717.

11. F. T. Arecchi, G. L. Masserini and P. Schwendimann, *Electromagnetic propagation in a resonant medium*, Riv. Nuovo Cimento **1** (1969), 181.

12. G. L. Lamb, Jr., *Analytical descriptions of ultrashort optical pulse propagation in a resonant medium*, Rev. Modern Phys. **43** (1971), 99.

13. J. Rubinstein, *Sine–Gordon equation*, J. Mathematical Phys. **11** (1970), 258–266. MR**41** #4958.

14. A. Barone, F. Esposito, C. J. Magee and A. C. Scott, *Theory and applications of the sine–Gordon equation*, Riv. Nuovo Cimento **1** (1971), 227.

15. J. K. Perring and T. H. R. Skyrme, *A model unified field equation*, Nuclear Phys. **31** (1962), 550–555. MR**25** #1840.

16. A. C. Scott, *Active and nonlinear wave propagation in electronics*, Wiley, New York, 1970, Chap. 5.

17. A. R. Forsyth, *Theory of differential equations*: 6. *Partial differential equations*, Dover, New York, 1959, p. 432. MR**23** # A1079.

18. A. C. Scott, *Propagation of magnetic flux on a long Josephson tunnel junction*, Nuovo Cimento **69B** (1970), 241.

19. D. W. McLaughlin and A. C. Scott, *A restricted Bäcklund transformation*, J. Math. Phys. (in press).

20. A. C. Scott, F. Y. F. Chu and D. W. McLaughlin, *The soliton*: *a new concept in applied science*, Proc. IEEE, October, 1973.

UNIVERSITY OF WISCONSIN

Lectures in Applied Mathematics
Volume 15, 1974

Numerical Solutions of the Korteweg–de Vries Equation and Its Generalizations by the Split–Step Fourier Method

Frederick Tappert

The split-step technique in conjunction with the fast Fourier transform algorithm has been found to provide an accurate, efficient, and stable numerical method for the solution of the following class of generalized Korteweg–de Vries equations:

$$u_t + f_x(u) + Lu_x = 0,$$

where f is a function of u, and L is a linear, constant coefficient, pseudo-differential operator, defined by its Fourier transform (denoted by $\mathscr{F}$),

$$\mathscr{F}(Lu) = S(k)\mathscr{F}(u),$$

where $S(k)$ is the symbol of L. The KdV equation corresponds to $f = u^2$, $S = a + bk^2$.

The essence of the solution method is to alternate between the following two steps: (1) advance the solution using only the nonlinear term by means of the implicit finite difference approximation,

$$\tilde{u}(t + \Delta t) = u(t) - \tfrac{1}{2}\Delta t[Q(\tilde{u}(t + \Delta t)) + Q(u(t))],$$

where $Q(u)$ approximates $f_x(u)$; (2) advance the solution exactly using only the linear term by means of the discrete fast Fourier transform,

$$u(t + \Delta t) = \mathscr{F}^{-1}(e^{ikS(k)\Delta t}\mathscr{F}(\tilde{u})).$$

AMS (*MOS*) *subject classifications* (1970). Primary 35-04, 35K55, 65M10; Secondary 35S10, 65M05.

Key words and phrases. Nonlinear partial differential equations, numerical methods for parabolic initial value problems.

This method is second order accurate and unconditionally stable (according to linear analysis).

The principal result of mathematical interest which has been inferred from an extensive (but not nearly exhaustive) series of numerical experiments may be stated tentatively as follows: the KdV equation is morphologically stable in the sense that an admixture of small perturbation terms, say $f(u) = u^2 + \varepsilon g(u)$ and $S(k) = a + bk^2 + \varepsilon T(k)$, does not appear to alter such qualitative features of solutions of the KdV equation as the recurrence of initial states and the nondisruptive interaction of solitons. This is rather surprising because it is known that if $g = u^4$ or $T = k^4$ (or $T = ik$) then the infinity of polynomial conservation laws is destroyed. Thus the existence of an infinity of polynomial conservation laws may be less significant than was previously supposed, and KdV type of behavior may be more general.

References

1. J. Y. T. Tang and F. D. Tappert, *Nonlinear interaction of finite amplitude waves* (unpublished; available from the second author).

2. F. D. Tappert and C. M. Varma, *Asymptotic theory of self-trapping of heat pulses in solids*, Phys. Rev. Lett. **25** (1970), 1108–1111. MR**42** #5526.

3. F. D. Tappert and R. H. Hardin, *New theoretical-numerical results on a model nonlinear dispersive wave equation*, SIAM Rev. (Chronicle) **13** (1971), 282.

4. F. D. Tappert, *The self-consistent wave kinetic equation*, SIAM Rev. (Chronicle) **13** (1971), 283.

5. F. D. Tappert and N. J. Zabusky, *Gradient-induced fission of solitons*, Phys. Rev. Lett. **27** (1971), 1774–1776.

6. F. D. Tappert, *Improved Korteweg–de Vries equation for ion acoustic waves*, Phys. Fluids **15** (1972), 2446–2447.

7. F. D. Tappert and C. N. Judice, *Recurrence of nonlinear ion acoustic waves*, Phys. Rev. Lett. **29** (1972), 1308–1311.

8. R. H. Hardin and F. D. Tappert, *Applications of the split-step Fourier method to the numerical solution of nonlinear and variable coefficient wave equations*, SIAM-SIGNUM Fall Meeting, Austin, Tex., October 1972; SIAM Rev. (Chronicle) **15** (1973), 423.

9. A. Hasegawa and F. Tappert, *Transmission of stationary nonlinear optical pulses in dispersive dielectric fibers*, Appl. Phys. Lett. **23** (1973), 171–172.

Bell Telephone Laboratories

Lectures in Applied Mathematics
Volume 15, 1974

Interfacial Solitary Waves in a Two-Fluid Medium

Lloyd R. Walker

Solitary waves are created at the interface between two immiscible liquids lying on a plane bed and having a free upper surface by means of a wavemaker programmed for variable displacement as a function of time. The experiment combines a model for geophysical flow problems with wave motion governed in first order by the Korteweg–de Vries equation. The predicted profile for this class of solutions is a hyperbolic secant squared with width proportional to the inverse square root of the amplitude, which is a function of the wave speed.

Due to tank size restrictions ($181 \times 54 \times 5$ cm), data are gathered for a small value of lower fluid depth ($h = 1$ cm) to allow an adequate (downstream distance)/(depth) ratio for assuring steady state waves. Surface tension is presumed to be significant on such scales, and existing theories are expanded to include this effect in the case of both free and rigid upper surface. Assuming static stability of the two-fluid medium at rest ($\rho =$ [upper density]/[lower density] < 1), and restricting experimentation to the gravity dominant regime ($h > h_{cr} = 0.53$ cm for the experimental parameters), the interfacial wave is of elevation if $r =$ (upper fluid depth)/(lower fluid depth) $\gg 1$ and of depression if $r \ll 1$. The critical value r_{cr} which divides the two alternatives is slightly greater than unity for the free surface, and slightly less than unity for the rigid lid cases, respectively, unaffected by surface tension. To first order terms in the normalized wave amplitude, α, the inclusion of capillary effects predicts unchanged wave

AMS (MOS) subject classifications (1970). Primary 76B25, 76C10; Secondary 76V05, 76D99, 86A05.

speed and narrowed wave shape, regardless of the upper boundary condition.

Experimental results show that the solitary waves created fit the width profile of $\operatorname{sech}^2 x$ to within 5 percent at all points, and that amplitude decay due to viscosity is accompanied by a broadening of the waveform, as predicted, with the fit to $\operatorname{sech}^2 x$ being retained. The half-widths of the wave are narrowed much more than would be expected on the basis of the surface tension correction alone. Speeds are found to be about 8 percent less than those predicted by the inviscid theory. The experimental value of r_{cr} is consistent with the theoretical prediction in the inviscid limit. Viscous effects are shown to offer a plausible explanation for the reduction in speed consistent with the concept of an interfacial boundary layer, in analogy to single fluid boundary layers, but exact analytical work is not done.

References

1. T. B. Benjamin, J. Fluid Mech. **25** (1966), 241.

2. ———, J. Fluid Mech. **29** (1967), 559.

3. G. H. Keulegan, J. Res. Nat. Bur. Standards **51** (1953), 133.

4. ———, Nat. Bur. Standards Report #4415, 1955.

5. D. J. Korteweg and G. de Vries, Philos. Mag. **39** (1895), 422.

6. P. D. Lax, Comm. Pure Appl. Math. **21** (1968), 467. MR**38** #3620.

7. O. S. Lee, Limn and Ocean **6** (1961), 312.

8. R. R. Long, Tellus **8** (1956), 460.

9. L. F. McGoldrick, Rev. Sci. Instr. **42** (1971), 359.

10. R. P. Mied and J. P. Dugan, Johns Hopkins University, Department of Mechanics Report, 1970 (unpublished).

11. R. L. Miller, in Proc. Sympos. Long Waves (University of Delaware, Newark, Del., Sept. 1970), 1972, p. 33.

12. A. S. Peters and J. J. Stoker, Comm. Pure Appl. Math. **13** (1960), 115. MR**22** #3296.

13. L. R. Walker, Phys. Fluids **16** (1973).

14. N. J. Zabusky and M. D. Kruskal, Phys. Rev. Lett. **15** (1965), 240.

University of Chicago

Bibliography of Recent Publications

The last year has seen an explosion in the amount of published material on the implementation of the inverse method for solving a wide class of nonlinear evolution equations and on the importance and application of the soliton in many areas of physics. In an effort to keep this volume as current as possible, we include below an extensive (but, I am sure, by no means complete) bibliography of recent and relevant publications. Some are not yet published and interested persons should write directly to the authors. Since the vast majority of these works are dated within the last year, it may seem incongruous to include references to the papers of Clairin (1903) and Loewner (1953). However, these works have not been widely read and deal with Backlund transformations which, some believe, may play a crucial role in the choice of the appropriate scattering problem for a given evolution equation. (It might be appropriate to mention in this context that the Ricatti equation and its properties also plays a central role.) I am indebted for the details of many of these references to the expository paper of Scott, Chu and McLaughlin ([**28**] below).

1. M. J. Ablowitz, D. J. Kaup, A. C. Newell and H. Segur, *The initial value solution for the sine–Gordon equation*, Phys. Rev. Lett. **30** (1973), 1262.

2. ———, *Nonlinear evolution equations of physical significance*, Phys. Rev. Lett. **31** (1973), 125.

3. M. J. Ablowitz and A. C. Newell, *The decay of continuous spectrum for solutions of the Korteweg–de Vries equation*, J. Mathematical Phys. **14** (1973), 1277.

4. M. J. Ablowitz, D. J. Kaup and A. C. Newell, *Self induced transparency, an irreversible, dispersive phenomenon* (to be published).

5. P. J. Caudrey, J. D. Gibbon, J. C. Eilbeck and R. K. Bullough, *Exact multisoliton solutions of the self induced transparency and sine–Gordon equations*, Phys. Rev. Lett. **30** (1973), 237–239.

6. M. J. Clairin, *Sur quelques équations aux dérivées partielles du second ordre*, Ann. Toulouse (5) **2** (1903), 437–458.

7. J. C. Eilbeck, J. D. Gibbon, P. J. Caudrey and R. K. Bullough, *Solitons in nonlinear optics*. I. *A more accurate description of the* 2π *pulse in self induced transparency* (to be published).

8. H. Flashka, *Integrability of the Toda lattice*. I Phys. Rev. (to appear).

9. ———, *Integrability of the Toda lattice*. II Prog. Theor. Phys. (to appear).

10. C. S. Gardner, J. M. Greene, M. D. Kruskal and R. M. Miura, *Korteweg–de Vries equation and generalizations*. VI. *Methods for exact solution*, 1973 (preprint).

11. J. L. Hammack, Jr., *Tsunamis–A model of their generation and propagation*, California Inst. Tech. Rep. KH-R-28, 1972, Chap. 5.

12. R. Hirota, *Exact solution of the Korteweg–de Vries equation for multiple collisions of solitons*, Phys. Rev. Lett. **27** (1971), 1192–1194.

13. ———, *Exact solution of the sine–Gordon equation for multiple collisions of solutions*, J. Phys. Soc. Japan **33** (1972), 1459–1463.

14. ———, *Exact N-soliton solutions of nonlinear lumped self-dual network equations*, J. Phys. Soc. Japan **35** (1973).

15. ———, *Exact N-soliton solutions of a nonlinear lumped network equation*, J. Phys. Soc. Japan **35** (1973).

16. ———, *Exact N-soliton solution of the wave equation of long waves in shallow water*, J. Mathematical Phys. (to be published).

17. ———, *Exact envelope–soliton solutions of a nonlinear wave equation*, J. Mathematical Phys. (to be published).

18. ———, *Exact solution of the modified Korteweg–de Vries equation for multiple collisions of solitons*, J. Phys. Soc. Japan **33** (1972), 1456–1458.

19. G. L. Lamb, Jr., M. O. Scully and F. A. Hopf, *Higher conservation laws of coherent optical pulse propagation in an inhomogeneously broadened medium*, Appl. Opl. **11** (1972), 2572–2575.

20. G. L. Lamb, Jr., *Coherent optical pulse propagation phase variation in*, Phys. Rev. Lett. **31** (1973), 196.

21. C. Loewner, *Generation of solutions of systems of partial differential equations by composition of infinitesimal Backlund transformations*, J. Analyse Math. **2** (1953), 219–242. MR **15**, 225.

22. S. L. McCall and E. L. Hahn, *Coherent light propagation through an inhomogeneously broadened 2-level system*, Bull. Amer. Phys. Soc. **10** (1965), 1189.

23. ———, *Self-induced transparency by pulsed coherent light*, Phys. Rev. Lett. **18** (1967), 908–911.

24. ———, *Self-induced transparency*, Phys. Rev. **183** (1969), 457–485.

25. D. W. McLaughlin and A. C. Scott, *A restricted Backlund transformation*, J. Mathematical Phys. (to be published).

26. K. Nozaki and T. Taniuti, *Propagation of solitary pulses in interaction of plasma waves*, J. Phys. Soc. Japan **34** (1973), 796–800.

27. M. Oikawa and N. Yajima, *Interactions of solitary waves—A perturbation approach to nonlinear systems*, J. Phys. Soc. Japan **34** (1973), 1093–1099.

28. A. C. Scott, F. Y. F. Chu and D. W. McLaughlin, *The soliton: A new concept in applied science*, Proc. IEEE, October, 1973.

29. A. B. Shabat, *A one-dimensional scattering theory*. I, Differential 'nye Uraunenija **8** (1972), 164–178. (Russian)

30. M. Toda and M. Wadati, *A soliton and two solitons in an exponential lattice and related equations*, J. Phys. Soc. Japan **34** (1973), 18–25.

31. M. Wadati and M. Toda, *The exact N-soliton solution of the Korteweg–de Vries equation*, J. Phys. Soc. Japan **32** (1972), 1403–1411.

32. M. Wadati, *The exact solution of the modified Korteweg–de Vries equation*, J. Phys. Soc. Japan **32** (1972), 1681.

33. ———, *The modified Korteweg–de Vries equation*, J. Phys. Soc. Japan **34** (1973), 1289–1296.

34. V. E. Zakharov and A. B. Shabat, *Exact theory of two-dimensional self-focusing and one-dimensional self-modulation of waves in nonlinear media*, Soviet Phys. JETP **34** (1972), 62–69.

Indexes

Author Addresses

Mark J. Ablowitz, Department of Mathematics, Clarkson College of Technology, Potsdam, New York 13676

T. Brooke Benjamin, Fluid Mechanics Research Institute, University of Essex, Colchester CO2 3SQ, England

D. J. Benney, Department of Mathematics, Massachusetts Institute of Technology, Cambridge, Massachusetts 02139

Norman Bleistein, Department of Mathematics, University of Denver, Denver, Colorado 80210

Jerry Bona, Department of Mathematics, University of Chicago, Chicago, Illinois 60637

Deb Bose, Fluid Mechanics Research Institute, University of Essex, Colchester CO2 3SQ, England

E. Fermi (Deceased)

James P. Fink, Department of Mathematics, University of Pittsburgh, Pittsburgh, Pennsylvania 15260

A. D. Gilbert, Fluid Mechanics Research Institute, University of Essex, Colchester CO2 3SQ, England

John M. Greene, Plasma Physics Laboratory, Princeton University, Princeton, New Jersey 08540

William S. Hall, Department of Mathematics, University of Pittsburgh, Pittsburgh, Pennsylvania 15260

George M. Homsy, Department of Chemical Engineering, Stanford University, Stanford, California 94305

Gray Jennings, 355 North Post Oak Lane, Apt. 643, Houston, Texas 77024

R. S. Johnson, School of Mathematics, University, Newcastle upon Tyne NE1 7R4, England

Siamak Khalili, Department of Mathematics, University of Pittsburgh, Pittsburgh, Pennsylvania 15260

Martin D. Kruskal, Peyton Hall, Princeton University, Princeton, New Jersey 08540

Peter D. Lax, Courant Institute of Mathematical Sciences, New York University, 251 Mercer Street, New York, New York 10021

Jon C. Luke, 2716 Mesa Verde Court, Burnsville, Minnesota 55337

J. J. Mahony, Department of Mathematics, University of Western Australia, Nedlands, W. A. 6009, Australia

Arthur Menikoff, Department of Mathematics, The Johns Hopkins University, Baltimore, Maryland 21218

Alan C. Newell, Department of Mathematics, Clarkson College of Technology, Potsdam, New York 13676

Takaaki Nishida, Department of Applied Mathematics and Physics, Faculty of Engineering, Kyoto University, Kyoto, Japan

John R. Pasta, Head, Office of Computing Activities, National Science Foundation, Washington, D.C. 20550

Alwyn C. Scott, Department of Electrical and Computer Engineering, University of Wisconsin, Madison, Wisconsin 53706

Harvey Segur, Department of Mathematics, Clarkson College of Technology, Potsdam, New York 13676

Ronald Smith, Fluid Mechanics Research Institute, University of Essex, Colchester CO4 3SQ, England

Joel A. Smoller, Department of Mathematics, University of Michigan, Ann Arbor, Michigan 48104

Frederick Tappert, Bell Laboratories, Whippany Road, Whippany, New Jersey 07981

S. Ulam, Department of Mathematics, University of Colorado, Boulder, Colorado 80302

Lloyd R. Walker, Shell Development Company, P.O. Box 481, Houston, Texas 77001

G. B. Whitham, Applied Mathematics, Firestone Laboratory, California Institute of Technology, Pasadena, California 91109

Referenced Authors

Roman numbers refer to pages on which a reference is made to an author or work of an author.
Italic numbers refer to pages on which a complete reference to a work by the author is given.
Boldface numbers indicate the first page of the articles in the book.